pages 2–3

pages 4–5

pages 6–7

CHEMISTRY OF MATTER

Anthea Maton
Former NSTA National Coordinator
Project Scope, Sequence, Coordination
Washington, DC

Jean Hopkins
Science Instructor and Department Chairperson
John H. Wood Middle School
San Antonio, Texas

Susan Johnson
Professor of Biology
Ball State University
Muncie, Indiana

David LaHart
Senior Instructor
Florida Solar Energy Center
Cape Canaveral, Florida

Charles William McLaughlin
Science Instructor and Department Chairperson
Central High School
St. Joseph, Missouri

Maryanna Quon Warner
Science Instructor
Del Dios Middle School
Escondido, California

Jill D. Wright
Professor of Science Education
Director of International Field Programs
University of Pittsburgh
Pittsburgh, Pennsylvania

 Prentice Hall
Englewood Cliffs, New Jersey
Needham, Massachusetts

Prentice Hall Science
Chemistry of Matter

Student Text and Annotated Teacher's Edition
Laboratory Manual
Teacher's Resource Package
Teacher's Desk Reference
Computer Test Bank
Teaching Transparencies
Product Testing Activities
Computer Courseware
Video and Interactive Video

The illustration on the cover, rendered by David Schleinkofer, shows a display of fireworks made possible by interactions of matter.

Credits begin on page 174.

SECOND EDITION

ISBN 0-13-400656-9

1 2 3 4 5 6 7 8 9 10   97 96 95 94 93

Prentice Hall
A Division of Simon & Schuster
Englewood Cliffs, New Jersey 07632

STAFF CREDITS

Editorial:	Harry Bakalian, Pamela E. Hirschfeld, Maureen Grassi, Robert P. Letendre, Elisa Mui Eiger, Lorraine Smith-Phelan, Christine A. Caputo
Design:	AnnMarie Roselli, Carmela Pereira, Susan Walrath, Leslie Osher, Art Soares
Production:	Suse F. Bell, Joan McCulley, Elizabeth Torjussen, Christina Burghard
Photo Research:	Libby Forsyth, Emily Rose, Martha Conway
Publishing Technology:	Andrew Grey Bommarito, Deborah Jones, Monduane Harris, Michael Colucci, Gregory Myers, Cleasta Wilburn
Marketing:	Andrew Socha, Victoria Willows
Pre-Press Production:	Laura Sanderson, Kathryn Dix, Denise Herckenrath
Manufacturing:	Rhett Conklin, Gertrude Szyferblatt

Consultants

Kathy French	National Science Consultant
Jeannie Dennard	National Science Consultant

CONTENTS

CHEMISTRY OF MATTER

Activity Bank/Reference Section

Features

CONCEPT MAPPING

Throughout your study of science, you will learn a variety of terms, facts, figures, and concepts. Each new topic you encounter will provide its own collection of words and ideas—which, at times, you may think seem endless. But each of the ideas within a particular topic is related in some way to the others. No concept in science is isolated. Thus it will help you to understand the topic if you see the whole picture; that is, the interconnectedness of all the individual terms and ideas. This is a much more effective and satisfying way of learning than memorizing separate facts.

Actually, this should be a rather familiar process for you. Although you may not think about it in this way, you analyze many of the elements in your daily life by looking for relationships or connections. For example, when you look at a collection of flowers, you may divide them into groups: roses, carnations, and daisies. You may then associate colors with these flowers: red, pink, and white. The general topic is flowers. The subtopic is types of flowers. And the colors are specific terms that describe flowers. A topic makes more sense and is more easily understood if you understand how it is broken down into individual ideas and how these ideas are related to one another and to the entire topic.

It is often helpful to organize information visually so that you can see how it all fits together. One technique for describing related ideas is called a **concept map**. In a concept map, an idea is represented by a word or phrase enclosed in a box. There are several ideas in any concept map. A connection between two ideas is made with a line. A word or two that describes the connection is written on or near the line. The general topic is located at the top of the map. That topic is then broken down into subtopics, or more specific ideas, by branching lines. The most specific topics are located at the bottom of the map.

To construct a concept map, first identify the important ideas or key terms in the chapter or section. Do not try to include too much information. Use your judgment as to what is really important. Write the general topic at the top of your map. Let's use an example to help illustrate this process. Suppose you decide that the key terms in a section you are reading are School, Living Things, Language Arts, Subtraction, Grammar, Mathematics, Experiments, Papers, Science, Addition, Novels. The general topic is School. Write and enclose this word in a box at the top of your map.

SCHOOL

Now choose the subtopics—Language Arts, Science, Mathematics. Figure out how they are related to the topic. Add these words to your map. Continue this procedure until you have included all the important ideas and terms. Then use lines to make the appropriate connections between ideas and terms. Don't forget to write a word or two on or near the connecting line to describe the nature of the connection.

Do not be concerned if you have to redraw your map (perhaps several times!) before you show all the important connections clearly. If, for example, you write papers for Science as well as for Language Arts, you may want to place these two subjects next to each other so that the lines do not overlap.

One more thing you should know about concept mapping: Concepts can be correctly mapped in many different ways. In fact, it is unlikely that any two people will draw identical concept maps for a complex topic. Thus there is no one correct concept map for any topic! Even

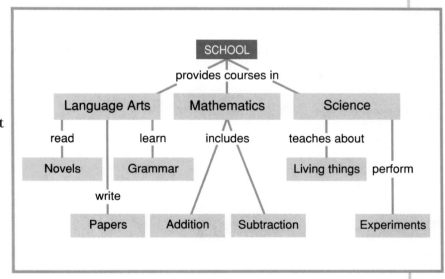

though your concept map may not match those of your classmates, it will be correct as long as it shows the most important concepts and the clear relationships among them. Your concept map will also be correct if it has meaning to you and if it helps you understand the material you are reading. A concept map should be so clear that if some of the terms are erased, the missing terms could easily be filled in by following the logic of the concept map.

CHEMISTRY OF MATTER

The fun of blowing bubbles is made possible by chemistry.

You have been given an assignment to perform the following steps, but you have not been told what the results of the procedure will be. What will you have made after completing the steps? Read them and see if you can figure it out. First, treat a fat called palmitin with an alkali such as sodium hydroxide in a process called saponification. The fat will break down to produce the substances sodium palmitate and glycerin. Discard the glycerin. Then add the sodium palmitate to a wetting compound to form a solution. Now dip a thin ring, preferably one with a handle, into the liquid. Finally, apply a gentle stream of air to the film on the ring.

Well, have you figured it out? Do you know what you have done? This rather complicated

The activities of the tiny particles that make up matter can be traced in patterns such as this one.

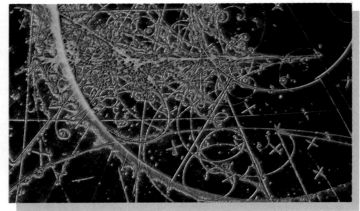

procedure has a fairly simple explanation: When you saponified the palmitin, you made soap. Then you added it to water to make a solution. And finally, you blew bubbles!

Chemical reactions such as saponification may not be familiar to you. Yet chemical reactions are occurring all around you and even in your body at this very moment. As you read this textbook, you will learn about the interactions of matter that can occur in a test tube, in nature, and even inside you!

Some of the most important chemical activities occur during the simple act of eating. ▶

Discovery *Activity*

Chemical Mysteries

1. Dip a toothpick into a small dish of milk. Use the toothpick as a pen to write a message on a sheet of white paper.

2. Let the milk dry. Observe what happens to your message as it dries.

3. Hold the paper close to a light bulb that is lit.

   ■ How does your message change before and after you place it next to the light bulb? What can you say about the relationship between the milk and the light bulb?

4. Fill one container about half full with very cold water and another container about half full with hot water.

5. Place four or five drops of food coloring into each container at exactly the same time.

   ■ Compare the rates at which the colors spread. Explain your observations.

Atoms and Bonding

Salt! You are surely familiar with this abundant and important chemical substance. As a frequently used seasoning, salt enhances the flavor of many of the foods you eat. Just imagine popcorn, chow mein, or collard greens without a dash of salt!

For a long time, salt has figured prominently in human affairs. Because it was both extremely important and limited in supply, salt was highly valued. Roman soldiers were, in fact, paid in cakes of salt. It is from the Latin word *sal,* meaning salt, that our word salary is derived.

Salt has come to symbolize many characteristics of human behavior. "The salt of youth" was William Shakespeare's description of the liveliness of this time of life. A person possessing valuable qualities is often described as being "worth one's salt." And those great human beings who have improved our world are often referred to as "the salt of the earth."

Salt is the substance sodium chloride. It is made of the elements sodium and chlorine. How and why do these elements combine to form salt? And what is the process by which hundreds of thousands of other substances are formed? Throw some salt over your left shoulder for luck and read on for the answers!

Journal *Activity*

You and Your World What do you think of when you hear the word element? How about the word compound? Do you know how these substances figure in your daily life? In your journal, answer these questions and provide any other thoughts you have about this topic. When you have completed this chapter, see if you need to modify your answers.

In ancient times, no other single substance equaled the importance of salt. Crystals of salt are shown in this magnified image.

1–1 What Is Chemical Bonding?

Look around you for a moment and describe what you see. Do you see the pages of this textbook? A window? Perhaps trees or buildings? Your friends and classmates? The air you breathe? All these ob-jects—and many others not even mentioned—have one important property in common. They are all forms of matter. And all matter—regardless of its size, shape, color, or phase—is made of tiny particles called **atoms**. Atoms are the basic building blocks of all the substances in the universe. As you can imag-ine, there are hundreds of thousands of different substances in nature. (To prove this, try to list all the different substances you can. Is there an end to your list?)

Yet, as scientists know, there are only 109 differ-ent elements. Elements are the simplest type of substance. Elements are made of only one kind of atom. How can just 109 different elements form so many different substances?

The 109 elements are each made of specific types of atoms. Atoms of elements combine with one an-other to produce new and different substances called compounds. You are already familiar with several compounds: water, sodium chloride (table salt), sugar, carbon dioxide, vinegar, lye, and ammonia. Compounds contain more than one kind of atom chemically joined together.

The combining of atoms of elements to form new substances is called chemical bonding. Chemical bonds are formed in very definite ways. The atoms combine according to certain rules. The rules of **chemical bonding** are determined by the structure of the atom, which you are now about to investigate.

Figure 1–1 *The hundreds of thousands of different substances in nature are made up of atoms. Atoms are the basic building blocks of everything in the universe, including all forms of living things, the air, and the Lagoon Nebula in the constellation Sagittarius.*

Electrons and Energy Levels

The atom contains a positively charged center called the nucleus. Found inside the nucleus are two types of subatomic (smaller than the atom) particles: protons and neutrons. Protons have a positive charge, and neutrons have no charge. Neutrons are

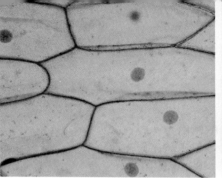
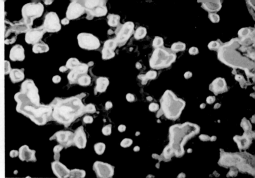

Figure 1–2 *You can think of atoms as similar to the individual stones that make up the pyramids in Giza, Egypt, or the individual cells of which all living things are made. The photograph of uranium atoms taken with an electron microscope gives you some idea of what atoms actually look like—magnified more than five million times, that is!*

neutral particles. Thus the nucleus as a whole has a positive charge.

Located outside the nucleus are negatively charged particles called electrons. The negative charge of the electrons balances the positive charge of the nucleus. The atom as a whole is neutral. It has no net charge. What do you think this means about the number of electrons (negatively charged) compared with the number of protons (positively charged)?

The negatively charged electrons of an atom are attracted by the positively charged nucleus of that atom. This electron-nucleus attraction holds the atom together. The electrons, however, are not pulled into the nucleus. They remain in a region outside the nucleus called the electron cloud.

The electron cloud is made up of a number of different energy levels. Electrons within an atom are arranged in energy levels. Each energy level can hold only a certain number of electrons. The first, or innermost, energy level can hold only 2 electrons, the second can hold 8 electrons, and the third can hold 18 electrons. The electrons in the outermost energy level of an atom are called **valence electrons.** The valence electrons play the most significant role in determining how atoms combine.

When the outermost energy level of an atom contains the maximum number of electrons, the level is full, or complete. Atoms that have complete (filled) outermost energy levels are very stable. They usually do not combine with other atoms to form compounds. They do not form chemical bonds.

ACTIVITY
DOING

A Model of Energy Levels

1. Cut a thin piece of corkboard into a circle 50 cm in diameter to represent an atom.

2. Insert a colored pushpin or tack into the center to represent the nucleus.

3. Draw three concentric circles around the nucleus to represent energy levels. The inner circle should be 20 cm in diameter; the second circle, 30 cm in diameter; and the third, 40 cm in diameter.

4. Using pushpins or tacks of another color to represent electrons, construct the following atoms: hydrogen (H), helium (He), lithium (Li), fluorine (F), neon (Ne), sodium (Na), and argon (Ar).

Are any of these elements in the same family? If so, which ones? How do you know?

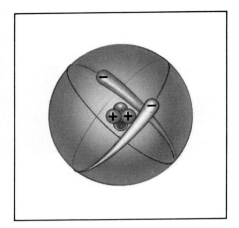

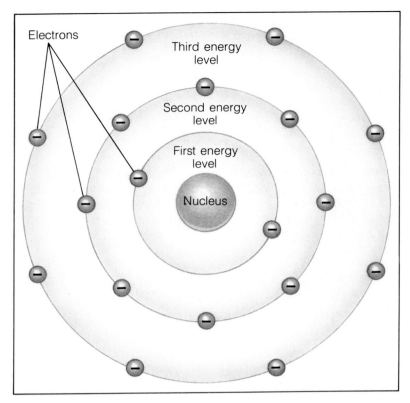

Figure 1–3 *An atom contains a positively charged nucleus surrounded by negatively charged electrons located in energy levels within the electron cloud. Each energy level can hold only a certain number of electrons. How many electrons can the first energy level hold? The second? The third?*

Up in Smoke, p.150

Turn to Appendix E on pages 166–167 and study it carefully. You are looking at the periodic table of the elements—one of the most important "tools" of a physical scientist. All the known elements (109) are listed in this table in a specific way. Every element belongs to a family, which is a numbered, vertical column. There are 18 families of elements. Every element also belongs to a period, which is a numbered, horizontal row. How many periods of elements do you see? Elements in the same family have similar properties, the most important of which is the number of electrons in the outermost energy level, or the number of valence electrons.

Look at Family 18 in the periodic table. It contains the elements helium, neon, argon, krypton, xenon, and radon. The atoms of these elements do not form chemical bonds under normal conditions. This is because all the atoms of elements in Family 18 have filled outermost energy levels. Remember, if the first energy level is also the outermost, it needs only 2 electrons to make it complete. Can you tell which element in Family 18 has only 2 valence electrons?

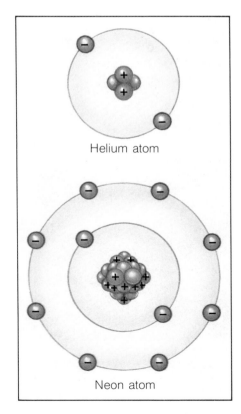

18

| 2
He
Helium
4.003 |
| 10
Ne
Neon
20.179 |
| 18
Ar
Argon
39.948 |
| 36
Kr
Krypton
83.80 |
| 54
Xe
Xenon
131.29 |
| 86
Rn
Radon
(222) |

Figure 1–4 *These balloons are filled with helium, a highly unreactive element. Neon gas, another unreactive element, is used in neon lights. Argon, shown here as a laser made visible through smoke, is also highly unreactive. These three elements are members of Family 18. What must be true of the outermost energy levels of atoms in this family?*

Electrons and Bonding

The electron arrangement of the outermost energy level of an atom determines whether or not the atom will form chemical bonds. As you have just read, atoms of elements in Family 18 have complete outermost energy levels. These atoms generally do not form chemical bonds.

Atoms of elements other than those in Family 18 do not have filled outermost energy levels. Their outermost energy level lacks one or more electrons to be complete. Some of these atoms tend to gain electrons in order to fill the outermost energy level. Fluorine (F), which has 7 valence electrons, gains 1 electron to fill its outermost energy level. Other atoms tend to lose their valence electrons and are left with only filled energy levels. Sodium (Na), which has 1 valence electron, loses 1 electron.

Helium atom

Neon atom

Figure 1–5 *The outermost energy level of a helium atom contains the maximum number of electrons—2. In a neon atom, the outermost energy level is the second energy level. It contains the maximum 8 electrons. What chemical property do these elements share?*

In order to achieve stability, an atom will either gain or lose electrons. In other words, an atom will bond with another atom if the bonding gives both atoms complete outermost energy levels. In the next section you will learn how bonding takes place.

1–1 Section Review

1. What is chemical bonding?
2. What is the basic structure of the atom?
3. What are valence electrons? How many valence electrons can there be in the first energy level? In the second? In the third?
4. What determines whether or not an atom will form chemical bonds?

Connection—*History*
5. On May 6, 1937, the airship (blimp) *Hindenburg* exploded in midair just seconds before completing its transatlantic voyage. On board the blimp had been 210,000 cubic meters of hydrogen gas. Since that time, airships have used helium gas rather than hydrogen gas to keep them aloft. Explain why.

Guide for Reading

Focus on this question as you read.

▶ *What is ionic bonding?*

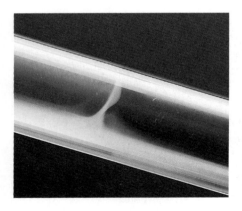

Figure 1–6 *Bonding usually results in the formation of compounds, such as ammonium chloride.*

1–2 Ionic Bonds

As you have just learned, an atom will bond with another atom in order to achieve stability, which means in order for both atoms to get complete outermost energy levels. One way a complete outermost energy level can be achieved is by the transfer of electrons from one atom to another. Bonding that involves a transfer of electrons is called **ionic bonding**. Ionic bonding, or electron-transfer bonding, gets its name from the word **ion**. An ion is a charged atom. Remember, an atom is neutral. But if there is a transfer of electrons, a neutral atom will become a charged atom.

Because ionic bonding involves the transfer of electrons, one atom gains electrons and the other atom loses electrons. Within each atom the negative and positive charges no longer balance. The atom that has gained electrons has gained a negative

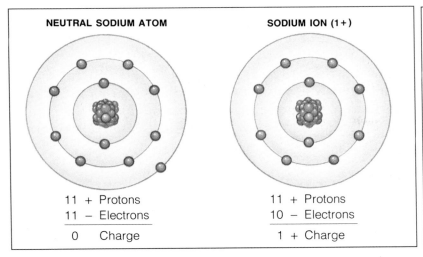

NEUTRAL SODIUM ATOM

11 + Protons
11 − Electrons
0   Charge

SODIUM ION (1+)

11 + Protons
10 − Electrons
1 + Charge

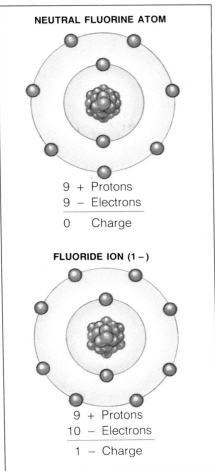

NEUTRAL FLUORINE ATOM

9 + Protons
9 − Electrons
0   Charge

FLUORIDE ION (1−)

9 + Protons
10 − Electrons
1 − Charge

charge. It is a negative ion. For example, fluorine (F) has 7 valence electrons. To complete its outermost energy level, the fluorine atom gains 1 electron. In gaining 1 negatively charged electron, the fluorine atom becomes a negative ion. The symbol for the fluoride ion is F^{1-}. (For certain elements, the name of the ion is slightly different from the name of the atom. The difference is usually in the ending of the name—as with the fluorine atom and the fluoride ion.)

The sodium atom (Na) has 1 valence electron. When a sodium atom loses this valence electron, it is left with an outermost energy level containing 8 electrons. In losing 1 negatively charged electron, the sodium atom becomes a positive ion. The symbol for the sodium ion is Na^{1+}. Figure 1–7 shows the formation of the fluoride ion and the sodium ion.

Figure 1–7 *The formation of a negative fluoride ion involves the gain of an electron by a fluorine atom. The formation of a positive sodium ion involves the loss of an electron by a sodium atom. How many valence electrons does a fluorine atom have? A sodium atom? What is the symbol for a fluoride ion? A sodium ion?*

Figure 1–8 *The general rule that opposites attract is responsible for the formation of the ionic bond between a positive sodium ion and a negative fluoride ion. Notice the transfer of an electron during the ionic bonding. What is the formula for the resulting compound?*

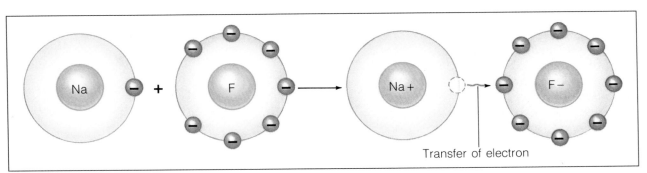

Transfer of electron

ACTIVITY
DISCOVERING

Steelwool Science

1. Place a small amount of steelwool in a jar and push it to the bottom. Pack it tightly enough so that it will remain in place when the jar is inverted.

2. Into each of two identical dishes, pour water to a height of about 2 cm. Make sure there are equal amounts of water in each dish.

3. In one dish, place the jar with the steelwool mouth down in the water. In the other dish, place an identical but empty jar in the same position.

4. Observe the jars every day for 1 week. Record your observations.

What changes did you observe in the steelwool? In the water levels?

■ Explain your observations.

■ What is the purpose of the empty jar?

In nature, it is a general rule that opposites attract. Since the two ions Na^{1+} and F^{1-} have opposite charges, they attract each other. The strong attraction holds the ions together in an ionic bond. The formation of the ionic bond results in the formation of the compound sodium fluoride, NaF. See Figure 1–8 on page 17.

Energy for Ion Formation

In order for the outermost electron to be removed from an atom, the attraction between the negatively charged electron and the positively charged nucleus must be overcome. The process of removing electrons and forming ions is called **ionization**. Energy is needed for ionization. This energy is called **ionization energy**.

The ionization energy for atoms that have few valence electrons is low. Do you know why? Only a small amount of energy is needed to remove electrons from the outermost energy level. As a result, these atoms tend to lose electrons easily and to become positive ions. What elements would you expect would have low ionization energies?

The ionization energy for atoms with many valence electrons is very high. These atoms do not lose electrons easily. As a matter of fact, these atoms usually gain electrons. It is much easier to gain 1 or 2 electrons than to lose 7 or 6 electrons! The tendency of an atom to attract electrons is called **electron affinity**. Atoms such as fluorine are said to have a high electron affinity because they attract electrons easily. What other atoms have a high electron affinity?

Figure 1–9 *During ionization, an electron is removed from an atom and an ion forms. Energy is absorbed during ionization. Energy is released when an atom gains an electron and forms an ion. What is the tendency of an atom to gain electrons called?*

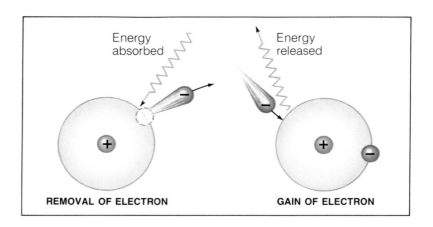

REMOVAL OF ELECTRON   GAIN OF ELECTRON

18 ■ O

Arrangement of Ions in Ionic Compounds

Ions of opposite charge strongly attract each other. Ions of like charge strongly repel each other. As a result, ions in an ionic compound are arranged in a specific way. Positive ions tend to be near negative ions and farther from other positive ions.

The placement of ions in an ionic compound results in a regular, repeating arrangement called a **crystal lattice**. A crystal lattice is made of huge numbers of ions. A crystal lattice gives the compound great stability. It also accounts for certain physical properties. For example, ionic solids tend to have high melting points. Figure 1–10 shows the crystal lattice structure of sodium chloride.

Ionic compounds are made of nearly endless arrays of ions. A chemical formula for an ionic compound shows the ratio of ions present in the crystal lattice. It does not show the actual number of ions.

Each ionic compound has a characteristic crystal lattice arrangement. This lattice arrangement gives a particular shape to the crystals of the compound. For example, sodium chloride forms cubic crystals. Figure 1–11 shows some unusual shapes that result from ionic bonding.

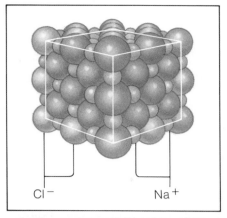

Figure 1–10 *Ionic bonding results in the formation of crystals. Crystals have a characteristic crystal lattice, or regular, repeating arrangement of ions. The lattice arrangement of sodium chloride crystals gives them their characteristic cubic shape. What is the common name for this crystal?*

Figure 1–11 *Ionic crystals often have unusual and amazingly beautiful shapes. Here you see ice crystals (left), crystals of the mineral rhodonite (center), and crystals of the mineral quartz (right).*

DOING

Growing Crystals

1. Make a small sliding loop in the end of a 10-cm length of thin plastic fishing line. Attach the loop to a small crystal of sea salt (NaCl).

2. In a beaker containing 200 mL of very hot water, stir to dissolve as much sea salt as the water will hold.

3. Suspend the fishing line containing the loop and crystal in the beaker. Tie the other end of the line to a pencil. Lay the pencil across the top of the beaker to support the line. The crystal should be suspended about halfway down in the liquid.

4. Observe the growing crystals every day for 3 days.

Guide for Reading

Focus on these questions as you read.

▶ *What is covalent bonding?*

▶ *What is a molecule?*

The crystal shape of an ionic compound is of great importance to geologists in identifying minerals. There are more than 2000 different kinds of minerals, and many of them look alike! One of the properties by which minerals are classified is crystal shape. There are six basic crystal shapes, or systems, and each of the thousands of minerals belongs to one of these systems.

1–2 Section Review

1. What is ionic bonding?
2. How does an atom become a negative ion? A positive ion?
3. What is ionization energy? Electron affinity?
4. What is a crystal lattice? What is the relationship between a chemical formula for an ionic compound and its crystal lattice?

Critical Thinking—*You and Your World*

5. A general rule in nature is that opposites attract. In addition to the behavior of oppositely charged ions, what other example(s) of this rule can you think of?

1–3 Covalent Bonds

Bonding often occurs between atoms that have high ionization energies and high electron affinities. In other words, neither atom loses electrons easily, but both atoms attract electrons. In such cases, there can be no transfer of electrons between atoms. What there can be is a sharing of electrons. Bonding in which electrons are shared rather than transferred is called **covalent bonding**. Look at the word covalent. Do you see a form of a word you have just learned? Do you know what the prefix *co-* means? Why is covalent an appropriate name for such a bond?

By sharing electrons, each atom fills up its outermost energy level. So the shared electrons are in the outermost energy level of both atoms at the same time.

Figure 1–12 *Covalent bonding is bonding in which electrons are shared rather than transferred. Two substances that exhibit covalent bonding are sulfur (left) and sugar (right). Which substance is an element? A compound?*

Nature of the Covalent Bond

In covalent bonding, the positively charged nucleus of each atom simultaneously (at the same time) **attracts the negatively charged electrons that are being shared.** The electrons spend most of their time between the atoms. The attraction between the nucleus and the shared electrons holds the atoms together.

The simplest kind of covalent bond is formed between two hydrogen atoms. Each hydrogen atom has 1 valence electron. By sharing their valence electrons, both hydrogen atoms fill their outermost energy level. Remember, the outermost energy level of a hydrogen atom is complete with 2 electrons. The two atoms are now joined in a covalent bond. See Figure 1–13 on page 22.

Chemists represent the electron sharing that takes place in a covalent bond by an **electron-dot diagram**. In such a diagram, the chemical symbol for an element represents the nucleus and all the inner energy levels of the atom—that is, all the energy levels except the outermost energy level, which is the energy level with the valence electrons. Dots surrounding the symbol represent the valence electrons.

A hydrogen atom has only 1 valence electron. An electron-dot diagram of a hydrogen atom would look like this:

H·

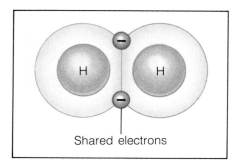

Shared electrons

Figure 1–13 *The covalent bond between 2 atoms of hydrogen results in a molecule of hydrogen. In a covalent bond, the electrons are shared. How many valence electrons does each hydrogen atom have?*

The covalent bond between the two hydrogen atoms shown in Figure 1–13 can be represented in an electron-dot diagram like this:

$$\text{H}:\text{H}$$

The two hydrogen atoms are sharing a pair of electrons. Each hydrogen atom achieves a complete outermost energy level (an energy level containing 2 electrons).

Chlorine has 7 valence electrons. An electron-dot diagram of a chlorine atom looks like this:

$$:\overset{\displaystyle\cdot\cdot}{\underset{\displaystyle\cdot\cdot}{\text{Cl}}}\cdot$$

The chlorine atom needs one more electron to complete its outermost energy level. If it bonds with another chlorine atom, the two atoms could share a pair of electrons. Each atom would then have 8 electrons in its outermost energy level. The electron-dot diagram for this covalent bond would look like this:

$$:\overset{\displaystyle\cdot\cdot}{\underset{\displaystyle\cdot\cdot}{\text{Cl}}}:\overset{\displaystyle\cdot\cdot}{\underset{\displaystyle\cdot\cdot}{\text{Cl}}}:$$

Covalent bonding often takes place between atoms of the same element. In addition to hydrogen and chlorine, the elements oxygen, fluorine, bromine, iodine, and nitrogen bond in this way. These elements are called **diatomic elements**. When found in nature, diatomic elements always exist as two atoms covalently bonded.

The chlorine atom, with its 7 valence electrons, can also bond covalently with an unlike atom. For example, a hydrogen atom can combine with a chlorine atom to form the compound hydrogen chloride. See Figure 1–14. The electron-dot diagram for this covalent bond is

$$\text{H}:\overset{\displaystyle\cdot\cdot}{\underset{\displaystyle\cdot\cdot}{\text{Cl}}}:$$

You can see from this electron-dot diagram that by sharing electrons, each atom completes its outermost energy level.

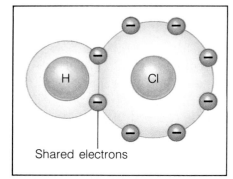

Shared electrons

Figure 1–14 *By sharing their valence electrons, hydrogen and chlorine form a molecule of the compound hydrogen chloride. When dissolved in water, hydrogen chloride forms the acid known as hydrochloric acid. Hydrochloric acid is an important part of the digestive juices found in the stomach.*

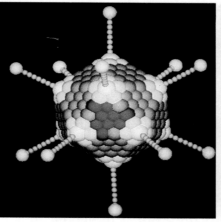

Figure 1–15 *Not all molecules are as simple as hydrogen chloride. Here you see computer-generated images of several more complex molecules: morphine (left), a common cold virus (center), and the hormone insulin (right).*

Formation of Molecules

In a covalent bond, a relatively small number of atoms are involved in the sharing of electrons. The combination of atoms that results forms a separate unit rather than the large crystal lattices characteristic of ionic compounds.

The combination of atoms formed by a covalent bond is called a **molecule** (MAHL-ih-kyool). A molecule is the smallest particle of a covalently bonded substance that has all the properties of that substance. This means that 1 molecule of water, for example, has all the characteristics of a glass of water, a bucket of water, or a pool of water. But if a molecule of water were broken down into atoms of its elements, the atoms would not have the same properties as the molecule.

Molecules are represented by chemical formulas. Like a chemical formula for an ionic crystal, the chemical formula for a covalent molecule contains the symbol of each element involved in the bond. Unlike a chemical formula for an ionic crystal, however, the chemical formula for a molecule shows the exact number of atoms of each element involved in the bond. The subscripts, or small numbers placed to the lower right of the symbols, show the number of atoms of each element. When there is only 1 atom of an element, the subscript 1 is not written. It is understood to be 1. Thus, a hydrogen chloride molecule has the formula HCl. What would be the

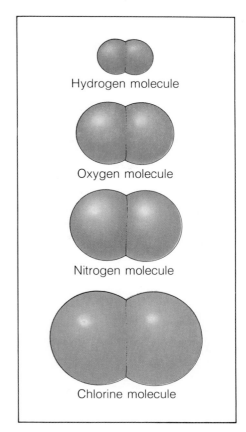

Figure 1–16 *Diatomic elements include hydrogen, oxygen, nitrogen, and chlorine. How many valence electrons does an atom of each element have?*

Figure 1–17 *Diamonds are network solids. Network solids contain bonds that are difficult to break. This accounts for the extreme hardness of diamonds. Extremely strong glues are also examples of network solids.*

formula for a molecule that has 1 carbon (C) atom and 4 chlorine (Cl) atoms?

Covalently bonded solids tend to have low melting points. Some covalent substances, however, do not have low melting points. They have rather high melting points. This is because molecules of these substances are very large. The molecules are large because the atoms involved continue to bond to one another. These substances are called **network solids**. Carbon in the form of graphite is an example of a network solid. So too is silicon dioxide, the main ingredient in sand. Certain glues also form networks of atoms whose bonds are difficult to break. This accounts for the holding properties of such glues.

Polyatomic Ions

Certain ions are made of covalently bonded atoms that tend to stay together as if they were a single atom. A group of covalently bonded atoms that acts like a single atom when combining with other atoms is called a **polyatomic ion**. Although the bonds within the polyatomic ion are covalent, the polyatomic ion usually forms ionic bonds with other atoms.

Figure 1–18 is a list of some of the more common polyatomic ions, and Figure 1–19 on page 26 shows the atomic structure of several polyatomic ions. Some of these ions may sound familiar to you. For example, the polyatomic ion hydrogen carbonate (HCO_3^{1-}) bonded to sodium produces sodium hydrogen carbonate ($NaHCO_3$), better known as baking soda. Magnesium hydroxide ($MgOH_2$) is milk of magnesia. And ammonium nitrate (NH_4NO_3—two polyatomic ions bonded together) is an important fertilizer.

POLYATOMIC IONS	
Name	**Formula**
ammonium	NH_4^{1+}
acetate	$C_2H_3O_2^{1-}$
chlorate	ClO_3^{1-}
hydrogen carbonate	HCO_3^{1-}
hydroxide	OH^{1-}
nitrate	NO_3^{1-}
nitrite	NO_2^{1-}
carbonate	CO_3^{2-}
sulfate	SO_4^{2-}
sulfite	SO_3^{2-}
phosphate	PO_4^{3-}

Figure 1–18 *The name and formula of some common polyatomic ions are shown here. What is a polyatomic ion? Which polyatomic ion has a positive charge?*

CONNECTIONS

Diamonds by Design

What do a razor blade, a rocket engine, and swamp gas have in common? Hardly anything, you might think. But today, researchers are working to provide an answer. Scientists involved in this endeavor are trying to find a way to coat various objects with a thin film of synthetic diamond. (Synthetic diamonds are made in the laboratory.)

Diamonds are extremely hard and resistant to wear. By placing a diamond coating on a razor blade, the blade can be made to last longer and stay sharper. Diamond coatings on rocket engines and cutting tools will increase their resistance to wear. The list of uses for diamond coatings goes on and on.

How do scientists go about making a synthetic diamond coating? They start with swamp gas, which is called methane. Methane is a chemical substance made of 1 carbon atom linked to 4 hydrogen atoms. The first step in the process is to "strip away" the hydrogen atoms from the carbon atom. When this happens, the carbon atoms from thousands of methane molecules are left behind.

Diamonds are made of carbon atoms. By carefully controlling conditions in the laboratory, scientists can make the carbon atoms link together to form synthetic diamonds. As they link together, they are deposited on the object to be coated. The synthetic diamonds are almost pure.

With continued research in this field of *chemical technology*, scientists feel certain that in any form, diamonds will continue to be extremely valuable!

Synthetic diamonds similar to these shown mixed with natural diamonds can be used to coat various objects, including silicon wafers, thereby making the objects stronger, sharper, and more durable.

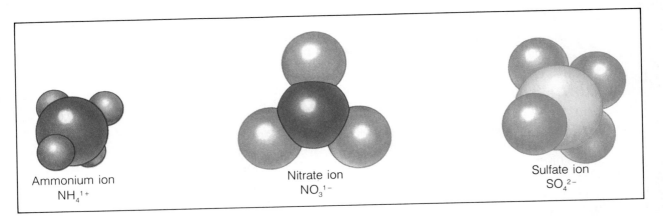

Ammonium ion
NH_4^{1+}

Nitrate ion
NO_3^{1-}

Sulfate ion
SO_4^{2-}

Figure 1–19 *A polyatomic ion is a group of covalently bonded atoms that act like a single atom when combining with other atoms. What kind of bond does a polyatomic ion usually form with another atom?*

1–3 Section Review

1. What is covalent bonding?
2. What is an electron-dot diagram? How is it used to represent a covalent bond?
3. What is a molecule? What does the chemical formula for a molecule tell you?
4. What is a polyatomic ion? Give two examples.

Critical Thinking—*Applying Concepts*
5. What elements and how many atoms of each are represented in the following formulas: Na_2CO_3, $Ca(OH)_2$, $Mg(C_2H_3O_2)_2$, $Ba_3(PO_4)_2$?

Guide for Reading

Focus on this question as you read.

▶ *What is a metallic bond?*

1–4 Metallic Bonds

You are probably familiar with metals such as copper, silver, gold, iron, tin, and zinc. And perhaps you even know that cadmium, nickel, chromium, and manganese are metals too. But do you know what makes an element a metal? Metals are elements that give up electrons easily.

In a metallic solid, or a solid made entirely of one metal element, only atoms of that particular metal are present. There are no other atoms to accept the electron(s) the metal easily gives up. How, then, do the atoms of a metal bond?

The atoms of metals form **metallic bonds**. In a metallic bond, the outer electrons of the atoms form

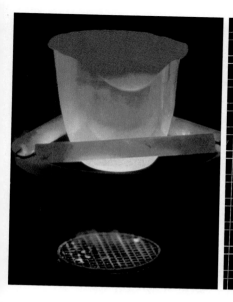

a common electron cloud. This common distribution of electrons occurs throughout a metallic crystal. In a sense, the electrons become the property of all the atoms. These electrons are often described as a "sea of electrons." **The positive nuclei of atoms of metals are surrounded by free-moving, or mobile, electrons that are all attracted by the nuclei at the same time.**

The sea of mobile electrons in a metallic crystal accounts for many properties of metals. Metals are malleable, which means they can be hammered into thin sheets without breaking. Metals are also ductile: They can be drawn into thin wire. The flexibility of metals results from the fact that the metal ions can slide by one another and the electrons are free to flow. Yet the attractions between the ions and the electrons hold the metal together even when it is being hammered or drawn into wire.

Figure 1–20 *The atoms of metals form metallic bonds. Metallic bonding accounts for many important properties of metals that make metals very useful. The metal platinum has an extremely high melting point, and so it is used in heat-resistant containers (left). The walls of this building are covered with a thin film of metal that reflects a significant amount of outdoor light (center). Metals are also excellent conductors of electricity. Some metals offer so little resistance to electric current that they can be used as superconductors (right).*

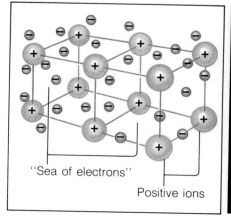

"Sea of electrons"

Positive ions

Figure 1–21 *In a metallic bond, the outer electrons of the metal atoms form a "sea of mobile electrons." Because metals are both malleable and ductile, they can be hammered and drawn into a wide variety of shapes. Here you see various forms of gold.*

The ability of the electrons to flow freely makes metals excellent conductors of both heat and electricity. Metallic bonding also accounts for the high melting point of most metals. For example, the melting point of silver is 961.9°C and of gold, 1064.4°C.

1–4 Section Review

1. What is a metallic bond?
2. What is a malleable metal? A ductile metal?
3. How does metallic bonding account for the properties of metals?

Connection—*You and Your World*
4. In terms of bonding, explain why it would be unwise to stir a hot liquid with a silver utensil.

Guide for Reading

Focus on these questions as you read.
▶ *What is oxidation number?*
▶ *What is the relationship between oxidation number and bond type?*

1–5 Predicting Types of Bonds

You have just learned about three different types of bonds formed between atoms of elements: ionic bonds, covalent bonds, and metallic bonds. By knowing some of the properties of an element, is there a way of predicting which type of bond it will form? Fortunately, the answer is yes. And the property most important for predicting bond type is the electron arrangement in the atoms of the element—more specifically, the number of valence electrons.

The placement of the elements involved in bonding in the periodic table often indicates whether the bond will be ionic, covalent, or metallic. Look again at the periodic table on pages 154–155. Elements at the left and in the center of the periodic table are metals. These elements have metallic bonds.

Compounds formed between elements that lose electrons easily and those that gain electrons easily will have ionic bonds. Elements at the left and in the center of the periodic table tend to lose valence electrons easily. These elements are metals. Elements at the right tend to gain electrons readily. These elements are nonmetals. A compound formed between a metal and a nonmetal will thus have ionic bonds.

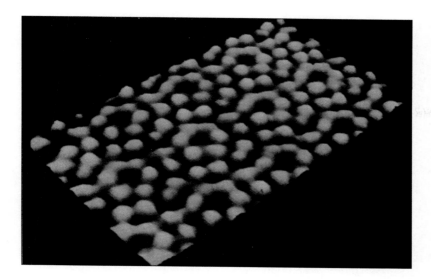

Figure 1–22 *This is the first photograph ever taken of atoms and their bonds. The bright round objects are single atoms. The fuzzy areas between atoms represent bonds.*

Compounds formed between elements that have similar tendencies to gain electrons will have covalent bonds. Bonds between nonmetals, which are at the right of the periodic table, will be covalent. What type of bonding would you expect between magnesium (Mg) and fluorine (F)? Between oxygen (O) and chlorine (Cl)? In a sample of zinc (Zn)?

Combining Capacity of Atoms

The number of electrons in the outermost energy level of an atom, the valence electrons, determines how an atom will combine with other atoms. If you know the number of valence electrons in an atom, you can calculate the number of electrons that atom needs to gain, lose, or share when it forms a compound. **The number of electrons an atom gains, loses, or shares when it forms chemical bonds is called its oxidation number.** The **oxidation number** of an atom describes its combining capacity.

An atom of sodium has 1 valence electron. It loses this electron when it combines with another atom. In so doing, it forms an ion with a 1+ charge, Na^{1+}. The oxidation number of sodium is 1+. A magnesium atom has 2 valence electrons, which it will lose when it forms a chemical bond. The magnesium ion is Mg^{2+}. The oxidation number of magnesium is 2+.

An atom of chlorine has 7 valence electrons. It will gain 1 electron when it bonds with another atom. The ion formed will have a 1− charge, Cl^{1-}. The oxidation number of chlorine is 1−. Oxygen has

Activity Bank

The Milky Way, p.152

ACTIVITY

CALCULATING

Determining Oxidation Numbers

For each of the following compounds, determine the oxidation number of the underlined element using the clue provided.

$\underline{Ca}Br_2$ (Br 1−)

$\underline{C}H_4$ (H 1−)

$\underline{H}_2S$ (S 2−)

$Na\underline{N}O_3$ (Na 1+, O 2−)

$\underline{Mg}SO_4$ ((SO$_4$) 2−)

Figure 1–23 *Some elements have more than one oxidation number. Copper, seen here as pennies, can have oxidation numbers of 1+ and 2+. Mercury also exhibits these same oxidation numbers. How can you determine the oxidation number of an atom?*

ACTIVITY

READING

Dangerous Food

Sometimes an element can have adverse effects on living things and on the environment. Such unwanted effects are often the result of bonding between the element and another substance. *Minamata,* a book by W. Eugene Smith and Aileen M. Smith, describes the tragic consequences to the Japanese population from the dumping of industrial mercury into Minamata Bay. You may find this story of the poisoning of a bay fascinating and important reading.

6 valence electrons. How many electrons will it gain? What is its oxidation number?

Using Oxidation Numbers

You can use the oxidation numbers of atoms to predict how atoms will combine and what the formula for the resulting compound will be. In order to do this, you must follow one important rule: *The sum of the oxidation numbers of the atoms in a compound must be zero.*

Sodium has an oxidation number of 1+. Chlorine has an oxidation number of 1−. One atom of sodium will bond with 1 atom of chlorine to form NaCl. Magnesium has an oxidation number of 2+. When magnesium bonds with chlorine, 1 atom of magnesium must combine with 2 atoms of chlorine, since each chlorine atom has an oxidation number of 1−. In other words, 2 atoms of chlorine are needed to gain the electrons lost by 1 atom of magnesium. The compound formed, magnesium chloride, contains 2 atoms of chlorine for each atom of magnesium. Its formula is $MgCl_2$. What would be the formula for calcium bromide? For sodium oxide? Remember the rule of oxidation numbers!

PROBLEM Solving

A Little Kitchen Chemistry

While preparing lunch one day in home economics class, Monduane and Juan noticed something unusual. As they added a dash of sugar to the vegetable oil they were using to make salad dressing, the sugar dissolved completely. But when they poured in a small amount of salt, as called for in the recipe they were following, the salt did not dissolve. Curious about their observation, the two chefs decided to repeat the process. They observed the same results. What, they wondered, could account for this difference in behavior? Would other substances show differences in their ability to dissolve in vegetable oil? In water?

Anxious to find the answers to their questions, the two students sought the advice of their teacher. The answer they received was simply this: Like dissolves in like.

Developing and Testing a Hypothesis

1. Help Monduane and Juan with their problem by suggesting a hypothesis.

2. Design an experiment to test your hypothesis. Remember to include a control.

3. What other applications does your hypothesis explain?

1–5 Section Review

1. What is an oxidation number?
2. How can the oxidation number of an atom be determined?
3. How is the oxidation number related to bond type?
4. What rule of oxidation numbers must be followed in writing chemical formulas?

Critical Thinking—*Making Predictions*
5. Predict the type of bond for each combination: Ca–Br, C–Cl, Ag–Ag, K–OH, SO_4^{2-}.

Laboratory Investigation

Properties of Ionic and Covalent Compounds

Problem

Do covalent compounds have different properties from ionic compounds?

Materials *(per group)*

safety goggles
salt
4 medium-sized
  test tubes
glass-marking
  pencil
test-tube tongs
Bunsen burner
timer
sugar
vegetable oil

distilled water
  (200 mL)
2 100-mL
  beakers
stirring rod
3 connecting
  wires
light bulb socket
light bulb
dry-cell battery

Procedure

1. Place a small sample of salt in a test tube. Label the test tube. Place an equal amount of sugar in another test tube. Label that test tube.

2. Using tongs, heat the test tube of salt over the flame of the Bunsen burner. **CAUTION:** *Observe all safety precautions when using a Bunsen burner.* Determine how long it takes for the salt to melt. Immediately stop heating when melting begins. Record the time.

3. Repeat step 2 using the sugar.

4. Half fill a test tube with vegetable oil. Place a small sample of salt in the test tube. Shake the test tube gently for about 10 seconds. Observe the results.

5. Repeat step 4 using the sugar.

6. Pour 50 mL of distilled water into a 100-mL beaker. Add some salt and stir until it is dissolved. To another 100-mL beaker add some sugar and stir until dissolved.

7. Using the beaker of salt water, set up a circuit as shown. **CAUTION:** *Exercise care when using electricity.* Observe the results. Repeat the procedure using the beaker of sugar water.

Observations

1. Does the salt or the sugar take a longer time to melt?

2. Does the salt dissolve in the vegetable oil? Does the sugar?

3. Which compound is a better conductor of electricity?

Analysis and Conclusions

1. Which substance do you think has a higher melting point? Explain.

2. Explain why one compound is a better conductor of electricity than the other.

3. How do the properties of each type of compound relate to their bonding?

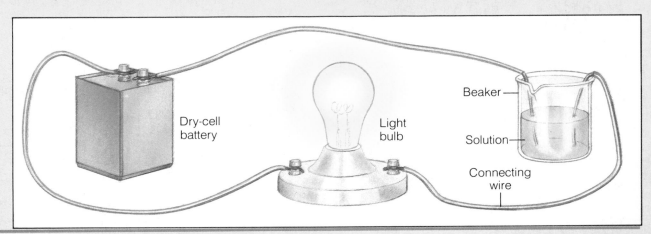

Dry-cell battery

Light bulb

Beaker

Solution

Connecting wire

Summarizing Key Concepts

1–1 What Is Chemical Bonding?

▲ All matter is made of tiny particles called atoms.

▲ Chemical bonding is the combining of elements to form new substances.

▲ The atom consists of a positively charged nucleus containing protons and neutrons, and energy levels containing electrons.

▲ Bonding involves the electrons in the outer-most energy level, the valence electrons.

1–2 Ionic Bonds

▲ Ionic bonding involves a transfer of electrons and a formation of ions.

▲ Ionization energy is the amount of energy needed to remove an electron from a neutral atom. Electron affinity is the tendency of an atom to attract electrons.

▲ The placement of ions in an ionic com-pound results in a crystal lattice.

1–3 Covalent Bonds

▲ Covalent bonding involves a sharing of electrons.

▲ A molecule is the smallest unit of a covalently bonded substance.

▲ Network solids are substances whose molecules are very large because the atoms in the substance continue to bond to one another.

▲ A polyatomic ion is a group of covalently bonded atoms that acts like a single atom when it combines with other atoms.

1–4 Metallic Bonds

▲ The basis of metallic bonding is the sea of mobile electrons that surrounds the nuclei and is simultaneously attracted by them.

1–5 Predicting Types of Bonds

▲ The oxidation number, or combining capacity, of an atom refers to the number of electrons the atom gains, loses, or shares when it forms chemical bonds.

▲ The oxidation number of any atom can be determined by knowing the number of elec-trons in its outermost energy level.

Reviewing Key Terms

Define each term in a complete sentence.

1–1 What Is Chemical Bonding?
atom
chemical bonding
valence electron

1–2 Ionic Bonds
ionic bonding
ion
ionization
ionization energy
electron affinity
crystal lattice

1–3 Covalent Bonds
covalent bonding
electron-dot diagram
diatomic element
molecule
network solid
polyatomic ion

1–4 Metallic Bonds
metallic bond

1–5 Predicting Types of Bonds
oxidation number

Chapter Review

Content Review

Multiple Choice

Choose the letter of the answer that best completes each statement.

1. Chemical bonding is the combining of elements to form new
   a. atoms.   c. substances.
   b. energy levels. d. electrons.
2. The center of an atom is called the
   a. electron.   c. octet.
   b. energy level.   d. nucleus.
3. The maximum number of electrons in the second energy level is
   a. 1.   b. 2.   c. 8.   d. 18.
4. Bonding that involves a transfer of electrons is called
   a. metallic.   c. ionic.
   b. covalent.   d. network.
5. Bonding that involves sharing of electrons within a molecule is called
   a. metallic.   c. covalent.
   b. ionic.   d. crystal.
6. The combination of atoms formed by covalent bonds is called a(an)
   a. element.   c. molecule.
   b. ion.   d. crystal.

7. Atoms that readily lose electrons have
   a. low ionization energy and low electron affinity.
   b. high ionization energy and low electron affinity.
   c. low ionization energy and high electron affinity.
   d. high ionization energy and high electron affinity.
8. An example of a polyatomic ion is
   a. SO_4^{2-} b. Ca^{2-} c. $NaCl$. d. O_2.
9. A sea of electrons is the basis of bonding in
   a. metals.
   b. ionic substances.
   c. nonmetals.
   d. covalent substances.
10. Bonding between atoms on the left and right sides of the periodic table tends to be
    a. covalent.   c. metallic.
    b. ionic.   d. impossible.

True or False

If the statement is true, write "true." If it is false, change the underlined word or words to make the statement true.

1. Electrons in the outermost energy level are called <u>oxidation</u> electrons.
2. <u>Covalent</u> bonds form crystals.
3. The tendency of an atom to attract electrons is called <u>electron affinity</u>.
4. The combining capacity of an atom is described by its <u>crystal lattice</u>.
5. <u>Malleable</u> solids are substances whose molecules are very large.
6. A charged atom is called a <u>molecule</u>.
7. Bromine is a <u>diatomic</u> element.

Concept Mapping

Complete the following concept map for Section 1–1. Refer to pages 06–07 to construct a concept map for the entire chapter.

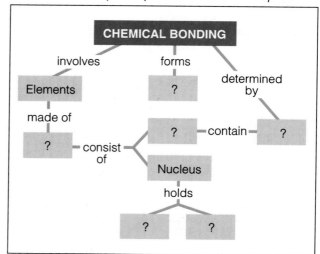

Concept Mastery

Discuss each of the following in a brief paragraph.

1. List the three types of chemical bonds and explain the differences among them.
2. Explain why the elements of Family 18 do not tend to form chemical bonds.
3. What are four properties of metals? How does the bonding in metals account for these properties?
4. How can you use the oxidation number of an atom to predict how it will bond?
5. Define the following structures that result from chemical bonds. Give one physical property of each.
   a. crystal lattice
   b. network solid
   c. covalently bonded solid
6. a. What is the difference between ionization energy and electron affinity?
   b. Why do atoms of high electron affinity tend to form ionic compounds with atoms of low ionization energy?
7. How can the periodic table be used to predict bond types?
8. Explain why the elements in Family 1 and in Family 17 are highly reactive.
9. How does chemical bonding account for the fact that although there are only 109 different elements, there are hundreds of thousands of different substances.
10. Describe the formation of a positive ion. Of a negative ion.

Critical Thinking and Problem Solving

Use the skills you have developed in this chapter to answer each of the following.

1. **Making predictions** Predict the type of bond formed by each pair of atoms. Explain your answers.
   a. Mg and Cl   c. ! and I
   b. Na and Na   d. Li and I
2. **Making diagrams** Draw the electron configuration for a Period 2 atom from each of the following families of the periodic table: Family 1, 2, 13, 14, 15, 16, 17, 18.
3. **Identifying patterns** Use the periodic table to predict the ion that each atom will form when bonding.
   a. sulfur (S)   d. astatine (At)
   b. rubidium (Rb)   e. sodium (Na)
   c. argon (Ar)   f. aluminum (Al)
4. **Applying facts** Draw an electron-dot diagram for the following molecules and explain why they are stable: F_2, NF_3.
5. **Making predictions** Use the periodic table to predict the formulas for the compounds formed by each of the following pairs of atoms.
   a. K and S   c. Ba and S
   b. Li and F   d. Mg and N
6. **Applying concepts** Explain why a sodium ion is smaller than a sodium atom.
7. **Identifying patterns** As you can see in the accompanying photograph, a sheet of aluminum metal is being cut. The ability of a metal to be cut into sections without shattering is called sectility. How can you account for the sectility of aluminum?

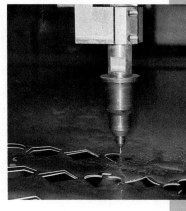

8. **Using the writing process** Pretend you are an experienced electron about to give an orientation lecture for "incoming freshmen electrons." Prepare your speech, being sure to include important details about placement in the atom, ionic bonding, covalent bonding, and metallic bonding.

Chemical Reactions

Guide for Reading

After you read the following sections, you will be able to

2–1 Nature of Chemical Reactions
- Describe the characteristics of chemical reactions.

2–2 Chemical Equations
- Write balanced chemical equations.

2–3 Types of Chemical Reactions
- Classify chemical reactions.

2–4 Energy of Chemical Reactions
- Describe energy changes in exothermic and endothermic reactions.

2–5 Rates of Chemical Reactions
- Apply the collision theory to factors that affect reaction rate.

Fireworks flash brilliantly in the night sky over the dark waters of the harbor. It is Independence Day. Amidst the wonderful celebration stands a very special lady. She towers above the waters, the torch in her upraised hand reaching high into the sky. She is a symbol of freedom, justice, and the brotherhood of people of all nations. Her name is Liberty.

She has stood there for over a century. But the passage of time had not been especially kind to her. The bronze of her outer structure, once bright and gleaming, had turned a dull gray-green. And the structure that supports her had begun to weaken. What had caused these changes? The answer has to do with the chemistry of atoms.

This chemistry, which damaged the Statue of Liberty, had also made possible the glorious restoration of this Lady in the Harbor. And the colorful fireworks lighting up the sky are products of the chemistry of atoms.

Chemical changes take place all the time, not just on the Fourth of July. And they take place everywhere, not just in New York Harbor. In this chapter you will learn about the nature of these chemical changes, many of which shape the world around you.

Journal *Activity*

You and Your World All around you, matter undergoes permanent changes. When bread is baked, for example, the dough changes forever. You can never turn the bread back into dough. A burned log can never be unburned. In your journal, describe several examples of substances that undergo changes and turn into new substances.

◀ *On July 4, 1986, fireworks lit up the sky in New York Harbor as the nation celebrated the one-hundredth birthday of the Statue of Liberty.*

2–1 Nature of Chemical Reactions

You have probably never given much thought to an ordinary book of matches. But stop for a minute and consider the fact that a single match in a book of matches can remain unchanged indefinitely. Yet if someone strikes that match, it bursts into a brilliant flame. And when that flame goes out, the appearance of the match will have changed forever. It can never be lighted again. The match has undergone a **chemical reaction.** What does this mean? **A chemical reaction is a process in which the physical and chemical properties of the original substances change as new substances with different physical and chemical properties are formed.** The burning of gasoline, the rusting of iron, and the baking of bread are all examples of chemical reactions.

Characteristics of Chemical Reactions

All chemical reactions share certain characteristics. One of these characteristics is that a chemical reaction always results in the formation of a new substance. The dark material on a burned match is a new substance. It is not the same substance that was originally on the match.

Another chemical reaction that you can easily observe occurs when a flashbulb lights. Because modern cameras have built-in flashes powered by a battery, you may not be familiar with traditional flashbulbs. At one time, however, all cameras used

Figure 2–1 *A burnt match has undergone a chemical reaction. So has rusted metal. Chemical reactions are also responsible for producing the vibrant colors of autumn leaves.*

flashbulbs similar to those shown in Figure 2–2 to provide the light necessary to take a photograph. Such flashbulbs can be used only once. You will now find out why.

Inside a flashbulb is a small coil of shiny gray metal. This metal is magnesium. The bulb is filled with the invisible gas oxygen. When the flashbulb is set off, the magnesium combines with the oxygen in a chemical reaction. During the reaction, energy is released in the form of light, and a fine white powder is produced. You can see this powder on the inside of the bulb. The powder is magnesium oxide, a compound with physical and chemical properties unlike those of the elements that were originally present—magnesium and oxygen. During the chemical reaction, the original substances are changed into a new substance. Now you can understand why traditional flashbulbs can be used only once.

The substances present before the change and the substances formed by the change are the two kinds of substances involved in a chemical reaction. A substance that enters into a chemical reaction is called a **reactant** (ree-AK-tuhnt). A substance that is produced by a chemical reaction is called a **product**. So a general description of a chemical reaction can be stated as reactants changing into products. In the example of the flashbulb, what are the reactants? The product?

In addition to changes in chemical and physical properties, chemical reactions always involve a

ACTIVITY

DISCOVERING

Flashes and Masses

1. Determine the mass of each of two unused flashbulbs. The masses should be the same.

2. Obtain a camera that uses flashbulbs. Put one flashbulb in the camera flash holder and flash the camera.

3. Allow the flashbulb to cool.

4. Again compare the masses of the used and the unused flashbulbs.

Does the mass of the used flashbulb change? What type of change has the used flashbulb undergone? How do you know?

■ What have you discovered about the mass of matter that undergoes a chemical reaction?

Figure 2–2 *Inside a flashbulb, oxygen surrounds a thin coil of magnesium. When the flashbulb is set off, a chemical reaction takes place in which magnesium combines with oxygen to form magnesium oxide. How can you tell a chemical reaction has occurred?*

change in energy. Energy is either absorbed or released during a chemical reaction. For example, heat energy is absorbed when sugar changes into caramel. When gasoline burns, heat energy is released. Later in this chapter you will learn more about the energy changes that accompany chemical reactions.

Capacity to React

In order for a chemical reaction to occur, the reactants must have the ability to combine with other substances to form products. What accounts for the ability of different substances to undergo certain chemical reactions? In order to answer this question, you must think back to what you learned about atoms and bonding.

Atoms contain electrons, or negatively charged particles. Electrons are located in energy levels surrounding the nucleus, or center of the atom. The electrons in the outermost energy level are called the valence electrons. It is the valence electrons that are involved in chemical bonding. An atom forms chemical bonds with other atoms in order to complete its outermost energy level. As you learned in Chapter 1, having a complete outermost energy level is the most stable condition for an atom. An atom will try to fill its outermost energy level by gaining or losing electrons, or by sharing electrons. A chemical bond formed by the gain or loss of electrons is an ionic bond. A chemical bond formed by the sharing of electrons is a covalent bond.

The arrangement of electrons in an atom determines the ease with which the atom will form chemical bonds. An atom whose outermost energy level is

Figure 2–3 *The chemical reactions that occur within sea coral regulate the amount of carbon dioxide in the ocean. Why do you think this is important?*

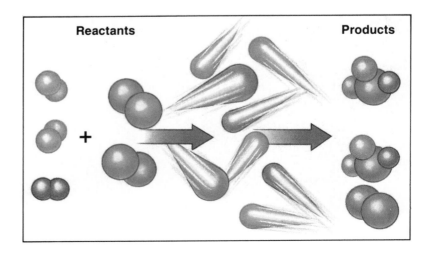

Figure 2–4 *During a chemical reaction, bonds between atoms of reactants are broken, atoms are rearranged, and new bonds in products are formed.*

Figure 2–5 *The tragic explosion of the Hindenburg on May 6, 1937, involved a chemical reaction. Hydrogen and oxygen combined explosively to form water. A much less dramatic chemical reaction is the corrosion of a metal such as copper, which was used to make the Statue of Liberty.*

full will not bond with other atoms. But an atom with an incomplete outermost energy level will bond readily. The ease with which an atom will form chemical bonds is known as the bonding capacity of an atom. The bonding capacity of an atom determines its ability to undergo chemical reactions. And the ability to undergo chemical reactions is an important chemical property.

During a chemical reaction, atoms can form molecules, molecules can break apart to form atoms, or molecules can react with other molecules. In any case, new substances are produced as existing bonds are broken, atoms are rearranged, and new bonds are formed.

2–1 Section Review

1. What is a chemical reaction?
2. What is a reactant? A product?
3. What is the relationship between the arrangement of electrons in an atom and the atom's chemical properties?

Critical Thinking—*Applying Concepts*
4. Cooking an egg until it is hard-boiled involves a chemical reaction. Cutting a piece of paper into a hundred little pieces does not involve a chemical reaction. Explain the difference between the two processes.

ACTIVITY READING

The Loss of the Hindenburg

The air was thick with excitement on May 6, 1937, as the German airship (blimp) *Hindenburg* headed toward its landing site in Lakehurst, New Jersey. The airship, held aloft by 210,000 cubic meters of hydrogen gas, was completing another transatlantic voyage. But before it could moor, the excitement turned to horror as the ship burst into flames. To learn about the *Hindenburg* disaster, read *Horror Overhead* by Richard A. Boning.

CAREERS

Food Chemist

Today, most methods of food processing involve chemical reactions. **Food chemists** use their knowledge of chemistry to develop these food processing methods.

Some food chemists develop new foods or new flavors. Others develop improved packaging and storage methods for foods.

If you are interested in a career as a food chemist, write to the Institute of Food Technology, Suite 300, 221 North LaSalle Street, Chicago, IL 60601.

2-2 Chemical Equations

It is important to be able to describe the details of a chemical reaction—how the reactants changed into the products. This involves indicating all the individual atoms involved in the reaction. One way of doing this is to use words. But describing a chemical reaction with words can be awkward. Many atoms may be involved, and the changes may be complicated.

For example, consider the flashbulb reaction described earlier. A word equation for this reaction would be: Magnesium combines with oxygen to form magnesium oxide and give off energy in the form of light. You could shorten this sentence by saying: Magnesium and oxygen form magnesium oxide and light energy.

Chemists have developed a more convenient way to represent a chemical reaction. Using symbols to represent elements and formulas to represent compounds, a chemical reaction can be described by a **chemical equation**. A chemical equation is an expression in which symbols and formulas are used to represent a chemical reaction.

In order to write a chemical equation, you must first write the correct chemical symbols or formulas for the reactants and products. Then you need to show that certain substances combine. This is done with the use of a + sign, which replaces the word and. Between the reactants and the products, you need to draw an arrow to show that the reactants have changed into the products. The arrow, which is read "yields," takes the place of an equal sign. It also shows the direction of the chemical change. The chemical equation for the flashbulb reaction can now be written:

$$Mg + O_2 \longrightarrow MgO + Energy$$

$$\text{Magnesium} + \text{Oxygen} \longrightarrow \text{Magnesium oxide} + \text{Energy}$$

Conservation of Mass

Chemists have long known that atoms can be neither created nor destroyed during a chemical reaction. In other words, the number of atoms of each

Figure 2–6 *Whether a chemical reaction is involved in the use of pesticides to protect crops, the violent eruption of a volcano, or the awesome explosion of a hydrogen bomb, one characteristic is always the same: Mass is never lost. What law describes this observation?*

element must be the same before and after the chemical reaction (that is, the number of atoms remains the same on both sides of the arrow in a chemical equation). The changes that occur during any chemical reaction involve only the rearrangement of atoms, not their production or destruction.

Every atom has a particular mass. Because the number of atoms of each element remains the same, mass can never change in a chemical reaction. The total mass of the reactants must equal the total mass of the products. No mass is lost or gained. **The observation that mass remains constant in a chemical reaction is known as the law of conservation of mass.**

Balancing Chemical Equations

The law of conservation of mass must be taken into account when writing a chemical equation for a chemical reaction. A chemical equation must show that atoms are neither created nor destroyed. The number of atoms of each element must be the same on both sides of the equation.

An equation in which the number of atoms of each element is the same on both sides of the equation is called a balanced chemical equation. Let's go back to the chemical equation for the flashbulb reaction:

$$Mg + O_2 \longrightarrow MgO + Energy$$

Is this a balanced chemical equation? Is the law of conservation of mass observed?

How many magnesium atoms do you count on the left side of the equation? You should count 1. And on the right side? You should count 1. Now try the same thing for oxygen. There are 2 oxygen atoms on the left but only 1 on the right. This cannot be correct because atoms can be neither created nor destroyed during a chemical reaction. How, then, can you make the number of atoms of each element the same on both sides of the equation?

One thing you cannot do to balance an equation is change a subscript. As you should remember from Chapter 1, a subscript is a small number placed to the lower right of a symbol. Changing the subscript would mean changing the substance. You can, however, change the number of atoms or molecules of each substance involved in the chemical reaction. You can do this by placing a number known as a coefficient (koh-uh-FIHSH-uhnt) in front of the appropriate symbols and formulas. Suppose the coefficient 3 is placed in front of a molecule of oxygen. It would be written as $3O_2$, and it would mean that there are 6 atoms of oxygen (3 molecules of 2 atoms each).

Now let's return to the flashbulb equation. To balance this equation, you must represent more than 1 atom of oxygen on the product side of the equation. If you place a coefficient of 2 in front of the formula for magnesium oxide, you will have 2 molecules of MgO. So you will have 2 atoms of oxygen. But you will also have 2 atoms of magnesium on the product side and only 1 atom of magnesium on the reactant side. So you must add a coefficient of 2 to the magnesium on the reactant side of the equation. There—the equation is balanced:

$$2\ Mg + O_2 \longrightarrow 2\ MgO + Energy$$

Figure 2–7 *These are the steps to follow in balancing a chemical equation. What law must a chemical equation obey?*

BALANCING EQUATIONS

1. Write a chemical equation with correct symbols and formulas.

$$H_2 + O_2 \rightarrow H_2O$$

2. Count the number of atoms of each element on each side of the arrow.

3. Balance atoms by using coefficients.

$$2H_2 + O_2 \rightarrow 2H_2O$$

4. Check your work by counting atoms of each element.

If you count atoms again, you will find 2 magnesium atoms on each side of the equation, as well as 2 oxygen atoms. The equation can be read: 2 atoms of magnesium combine with 1 molecule of oxygen to yield 2 molecules of magnesium oxide. Notice that when no coefficient is written, such as in front of the molecule of oxygen, the number is understood to be 1. Remember that to balance a chemical equation, you can change coefficients but never symbols or formulas.

Chemical equations are actually easy to write and balance. Follow the rules in Figure 2–7 and those listed here.

1. Write a word equation and then a chemical equation for the reaction. Make sure the symbols and formulas for reactants and products are correct.
2. Count the number of atoms of each element on each side of the arrow. If the numbers are the same, the equation is balanced.
3. If the number of atoms of each element is not the same on both sides of the arrow, you must balance the equation by using coefficients. Put a coefficient in front of a symbol or formula so that the number of atoms of that substance is the same on both sides of the arrow. Continue this procedure until you have balanced all the atoms.
4. Check your work by counting the atoms of each element to make sure they are the same on both sides of the equation.

ACTIVITY

CALCULATING

A Balancing Act

Rewrite each of the following equations on a sheet of paper and balance each.

$BaCl_2 + H_2SO_4 \rightarrow BaSO_4 + HCl$

$P + O_2 \rightarrow P_4O_{10}$

$KClO_3 \rightarrow KCl + O_2$

$C_3H_8 + O_2 \rightarrow CO_2 + H_2O$

$Cu + AgNO_3 \rightarrow Cu(NO_3)_2 + Ag$

2–2 Section Review

1. What is a chemical equation?
2. State the law of conservation of mass.
3. Why must a chemical equation be balanced?
4. Write a balanced chemical equation for the reaction between sodium and oxygen to form sodium oxide, Na_2O.
5. Why can't you change symbols, formulas, or subscripts in order to balance a chemical equation?

Connection—*Mathematics*
6. How are chemical equations related to equations found in mathematics?

PROBLEM ?·?·? Solving

The Silverware Mystery

The Smith family was about to sit down to a special dinner in honor of Mr. Smith's birthday when the telephone rang. The voice on the other end of the telephone informed Mrs. Smith that the family had won a contest. The prize was a fabulous trip that began immediately.

The family excitedly packed suitcases and prepared to leave the house. They wrapped their dinner and put it into the freezer. But they decided to leave the table as it was—beautifully arranged with flowers, china dishes, and silverware. They locked the doors and went on their way!

Two weeks later, after a wonderful trip, the Smiths returned. As they entered the dining room, they noticed something odd. The once shiny silver forks, knives, and spoons were now a dull, blackish-gray color. What could have happened? Could the silverware be returned to its original condition? Being a helpful neighbor, you decide to assist the Smiths in solving the silverware mystery in the following way.

- You tell them that it's all a matter of chemistry.
- You give them a rubber band, which contains sulfur, to experiment with.
- You ask them to think about what the air and the rubber band have in common.

Reaching Conclusions

1. What has happened to the Smiths' silverware?

2. How can it be returned to its original condition?

2–3 Types of Chemical Reactions

There are billions of different chemical reactions. In some reactions, elements combine to form compounds. In other reactions, compounds break down into elements. And in still other reactions, one element replaces another.

Chemists have identified four general types of reactions: synthesis, decomposition, single replacement, and double replacement. In each type of reaction, atoms are being rearranged and substances are being changed in a specific way.

Synthesis Reaction

In a **synthesis** (SIHN-thuh-sihs) **reaction**, two or more simple substances combine to form a new, more complex substance. So that you can easily identify synthesis reactions, it may be helpful for you to remember the form these reactions always take:

$$A + B \longrightarrow C$$

For example, the reaction between sodium and chlorine to form sodium chloride is a synthesis reaction:

$$2Na + Cl_2 \longrightarrow 2NaCl$$

Sodium + Chlorine $\longrightarrow$ Sodium chloride

Reactions involving the corrosion of metals are synthesis reactions. The rusting of iron involves the chemical combination of iron with oxygen to form iron oxide. Reactions in which a substance burns in oxygen are often synthesis reactions. Think back to the reaction that occurs in a flashbulb.

Decomposition Reaction

In a **decomposition reaction**, a complex substance breaks down into two or more simpler substances. Decomposition reactions are the reverse of synthesis reactions. Decomposition reactions take the form:

$$C \longrightarrow A + B$$

ACTIVITY

DISCOVERING

Preventing a Chemical Reaction

1. Obtain two large nails. Paint one nail and let it dry. Do not paint the other nail.

2. Pour a little water into a jar or beaker.

3. Stand both nails in the container of water. Cover the container and let it stand for several days. Compare the appearance of the nails.

Describe what happens to each nail. Give a reason for your observations. Write a word equation for any reaction that has occurred.

■ What effect does paint have on the rusting process of a nail?

Figure 2–8 *A reaction in which a substance burns in oxygen is a synthesis reaction. Why do you think smoking is not permitted in areas where oxygen is being administered?*

Figure 2–9 *Carbonic acid gives liquids their "fizz." Carbonic acid, however, quickly decomposes into water and carbon dioxide gas. The reaction that occurs when explosives are ignited is also a decomposition reaction.*

Popcorn Hop, p.154

ACTIVITY
DOING

The Disappearing Coin

1. Place a small piece of aluminum foil in a glass filled with water.

2. Position a copper coin, such as a penny, on top of the foil.

3. Let the glass stand for one day and observe what happens.

Describe the appearance of the water at the end of the experiment. Of the aluminum foil. Of the coin.

What chemical reaction has taken place? How do you know?

When you take the cap off a bottle of soda, bubbles rise quickly to the top. Why? Carbonated beverages such as soda contain the compound carbonic acid, H_2CO_3. This compound decomposes into water (H_2O) and carbon dioxide gas (CO_2). The CO_2 gas makes up the bubbles that are released. The balanced equation for the decomposition of carbonic acid is

$$H_2CO_3 \longrightarrow H_2O + CO_2$$

Carbonic acid ⟶ Water + Carbon dioxide

Single-Replacement Reaction

In a **single-replacement reaction**, an uncombined element replaces an element that is part of a compound. These reactions take the form:

$$A + BX \longrightarrow AX + B$$

Notice that the atom represented by the letter X switches its partner from B to A.

An example of a single-replacement reaction is the reaction between sodium and water. The very active metal sodium must be stored in oil, not water. When it comes in contact with water, it reacts explosively. The sodium replaces the hydrogen in the water and releases lots of energy. The balanced equation for the reaction of sodium with water is:

$$2Na + 2H_2O \longrightarrow 2NaOH + H_2$$

Sodium + Water ⟶ Sodium hydroxide + Hydrogen

Most single-replacement reactions, however, do not cause explosions.

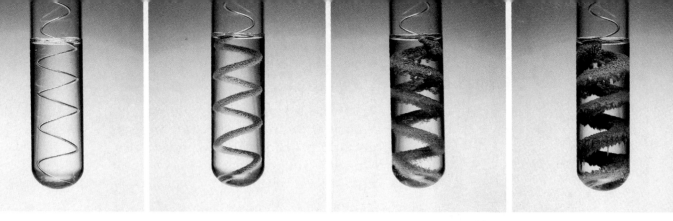

Double-Replacement Reaction

In a **double-replacement reaction**, different atoms in two different compounds replace each other. In other words, two compounds react to form two new compounds. These reactions take the form:

$$AX + BY \longrightarrow AY + BX$$

Notice that in this reaction the atoms represented by both the letters X and Y switch partners.

If you have ever had an upset stomach, you may have taken a medicine that contained the compound magnesium carbonate. This compound reacts with the hydrochloric acid in your stomach in the following way:

$$MgCO_3 + 2\ HCl \longrightarrow MgCl_2 + H_2CO_3$$

Magnesium carbonate + Hydrochloric acid $\longrightarrow$ **Magnesium chloride + Carbonic acid**

In this double-replacement reaction, the magnesium and hydrogen replace each other, or switch partners. One product is magnesium chloride, a harmless compound. The other product is carbonic acid. Do you remember what happens to carbonic acid? It decomposes into water and carbon dioxide. Your stomachache goes away because instead of too much acid, there is now water and carbon dioxide. You owe your relief to this double-replacement reaction:

$$MgCO_3 + 2HCl \longrightarrow MgCl_2 + H_2O + CO_2$$

Magnesium carbonate + Hydrochloric acid $\longrightarrow$ **Magnesium chloride + Water + Carbon dioxide**

Figure 2–10 *Because copper is a more active metal than silver, it can replace the silver in silver nitrate. In these four photos, you can see the gradual buildup of silver metal on the coil. What type of reaction is this? What other indication is there that a chemical change is taking place?*

ACTIVITY

DISCOVERING

Double-Replacement Reaction

1. Place a small amount of baking soda in a glass beaker or jar.

2. Pour some vinegar on the baking soda. Observe what happens.

Baking soda is sodium hydrogen carbonate, $NaHCO_3$. Vinegar is acetic acid, $HC_2H_3O_2$. Write the chemical equation for this reaction. What gas is produced?

■ How could you test for the presence of this gas?

Figure 2–11 *Paints are chemical compounds produced by double-replacement reactions. What is the general form of a double-replacement reaction?*

2–3 Section Review

1. Name the four types of reactions.
2. What is the difference between a synthesis reaction and a decomposition reaction?
3. What is a single-replacement reaction? A double-replacement reaction?

Critical Thinking—*Identifying Reactions*

4. What type of reaction is represented by each of the following equations:
   a. $CaCO_3 \longrightarrow CaO + CO_2$
   b. $C + O_2 \longrightarrow CO_2$
   c. $BaBr_2 + K_2SO_4 \longrightarrow 2KBr + BaSO_4$

Guide for Reading

Focus on this question as you read.

▶ *How are chemical reactions classified according to energy changes?*

Figure 2–12 *The explosion of a firecracker is an exothermic reaction. Why is this one reason that firecrackers are dangerous?*

2–4 Energy of Chemical Reactions

Energy is always involved in a chemical reaction. Sometimes energy is released, or given off, as the reaction takes place. Sometimes energy is absorbed. **Based on the type of energy change involved, chemical reactions are classified as either exothermic or endothermic reactions.**

In either type of reaction, energy is neither created nor destroyed. It merely changes position or form. The energy released or absorbed usually takes the form of heat or visible light.

Exothermic Reactions

A chemical reaction in which energy is released is an **exothermic** (ek-soh-THER-mihk) **reaction.** The word exothermic comes from the root -*thermic,* which refers to heat, and the prefix *exo-*, which means out of. Heat comes out of, or is released from, a reacting substance during an exothermic reaction. A reaction that involves burning, or a combustion reaction, is an example of an exothermic reaction. The combustion of methane gas, which occurs in a gas stove, releases a large amount of heat energy.

The energy that is released in an exothermic reaction was originally stored in the molecules of the reactants. Because the energy is released during the reaction, the molecules of the products do not receive this energy. So the energy of the products is less than the energy of the reactants. Energy diagrams, such as the ones shown in Figure 2–14, can be used to show the energy change in a reaction. Note that in an exothermic reaction, the reactants are higher in energy than the products are.

Endothermic Reactions

A chemical reaction in which energy is absorbed is an **endothermic reaction**. The prefix *endo-* means into. During an endothermic reaction, energy is taken into a reacting substance. The energy absorbed during an endothermic reaction is usually in the form of heat or light. The decomposition of sodium chloride, or table salt, is an endothermic reaction. It requires the absorption of electric energy.

The energy that is absorbed in an endothermic reaction is now stored in the molecules of the products. So the energy of the products is more than the energy of the reactants. See Figure 2–14 again.

Activation Energy

The total energy released or absorbed by a chemical reaction does not tell the whole story about the energy changes involved in the reaction. In order for the reactants to form products, the molecules of the reactants must combine to form a short-lived, high-energy, extremely unstable molecule. The atoms of this molecule are then rearranged to form products. This process requires energy. The molecules of

Figure 2-13 *The cooking of pancakes is an endothermic reaction. Why?*

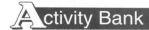

Activity Bank

Toasting to Good Health, p.155

Figure 2–14 *An energy diagram for an exothermic reaction indicates that heat is released during the reaction. Heat is absorbed during an endothermic reaction, as shown by its energy diagram. How does the heat content of products and reactants compare for each type of reaction?*

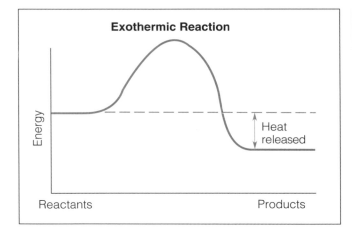

Exothermic Reaction

Energy

Heat released

Reactants   Products

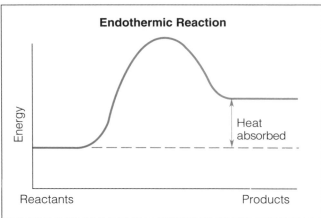

Endothermic Reaction

Energy

Heat absorbed

Reactants   Products

Figure 2–15 *As you can see by these energy diagrams, both an exothermic reaction and an endothermic reaction require activation energy.*

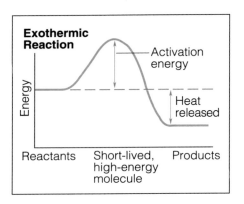

Exothermic Reaction

Activation energy

Heat released

Energy

Reactants | Short-lived, high-energy molecule | Products

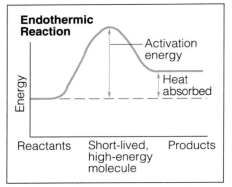

Endothermic Reaction

Activation energy

Heat absorbed

Energy

Reactants | Short-lived, high-energy molecule | Products

the reactants must "climb" to the top of an "energy hill" before they can form products. The energy needed to "climb" to the top of the "energy hill" is called **activation energy**. After the reactants have absorbed this activation energy, they can "slide down" the energy hill to form products.

An energy diagram indicates more than whether a reaction is exothermic or endothermic. An energy diagram shows the activation energy of the reaction. Figure 2–15 shows an energy diagram for both an exothermic reaction and an endothermic reaction.

All chemical reactions require activation energy. Even an exothermic reaction such as the burning of a match requires activation energy. In order to light a match, it must first be struck. The friction of the match against the striking pad provides the necessary activation energy.

ACTIVITY
THINKING

Kitchen Chemistry

Many interesting chemical reactions occur during various cooking processes. Observe someone preparing and cooking different kinds of food. Record your observations of the changes that take place in the properties of the food. Are the changes physical or chemical? Endothermic or exothermic? Synthesis, decomposition, or replacement?

Using books and other reference materials in the library, find out more about each of the chemical reactions you have listed.

2–4 Section Review

1. What is an exothermic reaction? An endothermic reaction?
2. On which side should the energy term be written in an equation representing an endothermic reaction? In an equation representing an exothermic reaction?
3. Compare the energy content of reactants and products in an exothermic reaction. In an endothermic reaction.

Connection—*You and Your World*
4. Using what you know about activation energy, explain why a match will not light if it is not struck hard enough.

2–5 Rates of Chemical Reactions

Guide for Reading

Focus on this question as you read.

▶ *How is the collision theory related to the factors affecting reaction rate?*

The complete burning of a thick log can take many hours. Yet if the log is ground into fine sawdust, the burning can take place at dangerously high speeds. In fact, if the dust is spread through the air, the burning can produce an explosion! In both these processes, the same reaction is taking place. The various substances in the wood are combining with oxygen. One reaction, however, proceeds at a faster speed than the other one does. What causes the differences in reaction times?

In order to explain differences in reaction time, chemists must study **kinetics**. Kinetics is the study of **reaction rates**. The rate of a reaction is a measure of how quickly reactants turn into products. Reaction rates depend on a number of factors, which you will now read about.

Collision Theory

You learned that chemical reactions occur when bonds between atoms are broken, the atoms are rearranged, and new bonds are formed. In order for this process to occur, particles must collide. As two particles approach each other, they begin to interact. During this interaction, old bonds may be broken and new bonds formed. For a reaction to occur, however, particles must collide at precisely the correct angle with the proper amount of energy. The more collisions that occur under these conditions, the faster the rate of the chemical reaction.

A theory known as the **collision theory** relates particle collisions to reaction rate. **According to the collision theory, the rate of a reaction is affected by four factors: concentration, surface area, temperature, and catalysts.**

Concentration

The concentration of a substance is a measure of the amount of that substance in a given unit of volume. A high concentration of reactants means there

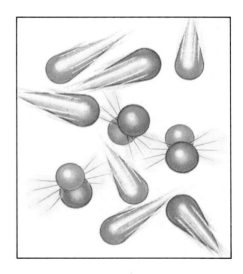

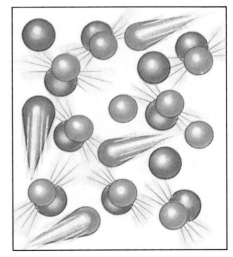

Figure 2–16 *Collisions of molecules increase when there are more molecules. The number of molecules per unit volume is called concentration. What is the relationship between the concentration of reactants and reaction rate?*

Figure 2–17 *As bellows are pumped, more oxygen is supplied to the fire and the rate of reaction increases. Why are forest fires particularly dangerous when weather conditions nearby include high winds?*

are a great many particles per unit volume. So there are more particles of reactants available for collisions. More collisions occur and more products are formed in a certain amount of time. What does a low concentration of reactants mean?

Generally, most chemical reactions proceed at a faster rate if the concentration of the reactants is increased. A decrease in the concentration of the reactants decreases the rate of reaction. For example, a highly concentrated solution of sodium hydroxide (NaOH), or lye, will react more quickly to clear a clogged drain than will a less concentrated lye solution. Why would the rate of burning charcoal be increased by blowing on the fire?

Surface Area

When one of the reactants in a chemical reaction is a solid, the rate of reaction can be increased by breaking the solid into smaller pieces. This increases the surface area of the reactant. Surface area refers to how much of a material is exposed. An increase in surface area increases the collisions between reacting particles.

A given quantity of wood burns faster as sawdust than it does as logs. Sawdust has a much greater surface area exposed to air than do the logs. So oxygen particles from the air can collide with more wood particles per second. The reaction rate is increased.

Figure 2–18 *It may seem hard to believe that grains of wheat or even dust can cause the immense destruction you see here. But an explosion at a grain elevator is an ever-present danger because a chemical reaction can occur almost instantaneously. What reaction-rate factor is responsible for such an explosion?*

Many medicines are produced in the form of a fine powder or many small crystals. Medicine in this form is often more effective than the same medicine in tablet form. Do you know why? Tablets dissolve in the stomach and enter the bloodstream at a slower rate. How does the collision theory account for the fact that fine crystals of table salt dissolve more quickly in water than do large crystals of rock salt?

Temperature

An increase in temperature generally increases the rate of a reaction. Here again the collision theory provides an explanation for this fact. Particles are constantly in motion. Temperature is a measure of the energy of their motion. Particles at a high temperature have more energy of motion than do particles at a low temperature. Particles at a high temperature move faster than do particles at a low temperature. So particles at a high temperature collide more frequently. They also collide with greater energy. This increase in the rate and energy of collisions affects the reaction rate. More particles of reactants are able to gain the activation energy needed to form products. So reaction rate is increased.

At room temperature, the rates of many chemical reactions roughly double or triple with a rise in temperature of 10°C. How does this fact explain the use of refrigeration to keep foods from spoiling?

Catalysts

Some chemical reactions take place very slowly. The reactions involved with digesting a cookie are examples. In fact, if these reactions proceeded at their normal rate, it could take weeks to digest one cookie! Fortunately, certain substances speed up the rate of a chemical reaction. These substances are **catalysts**. A catalyst is a substance that increases the rate of a reaction but is not itself changed by the reaction. Although a catalyst alters the reaction, it can be recovered at the end of the reaction.

How does a catalyst change the rate of a reaction if it is not changed by the reaction? Again, the explanation is based on the collision theory. Reactions often involve a series of steps. A catalyst changes one

Figure 2–19 *These glow-in-the-dark sticks are called Cyalume light sticks. When a light stick is placed in hot water (left), it glows more brightly than when placed in cold water (right). The reaction that causes a stick to glow is faster and produces more light at higher temperatures.*

ACTIVITY
DISCOVERING

Temperature and Reaction Rate

1. Fill one glass with cold water and another with hot water.

2. Drop a seltzer tablet into each glass of water and observe the reactions that occur.

Is there any noticeable difference in the two reactions?

■ What relationship regarding reaction rate does this activity illustrate?

■ Does this activity prove that a temperature difference always has the same effect?

Figure 2–20 *A catalyst changes the rate of a chemical reaction without itself being changed by the reaction. According to this energy diagram, how does a catalyst affect the rate of a reaction?*

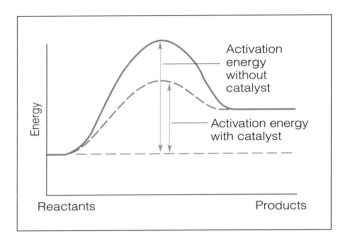

Figure 2–21 *This collection of colorful pellets actually contains catalysts used in industry. The blue catalyst in front, for example, is used in the reactions that remove sulfur and nitrogen from crude oil.*

or more of the steps. A catalyst produces a different, lower energy path for the reaction. In other words, it lowers the "energy hill," or activation energy. A decrease in the activation energy needed for the reaction allows more reactant particles to form products. Collisions need not be so energetic. Therefore, more collisions are successful at producing products.

A catalyst is usually involved in one or more of the early steps in a reaction. The catalyst is, however, re-formed during a later step. This explains why a catalyst can be recovered at the end of the reaction.

Catalysts are used in many chemical processes. Certain automobiles contain devices called catalytic converters. A catalytic converter speeds up the reaction that changes the harmful gases produced by automobile engines into harmless ones. Some of the most important catalysts are those found in your body. Catalysts in the body are called enzymes. Each enzyme increases the rate of a specific reaction involved in the body's metabolism.

2–5 Section Review

1. What is reaction rate?
2. How is reaction rate related to collision theory?
3. Name four factors that affect reaction rate.
4. How does collision theory explain the effect of a decrease in temperature on the reaction rate?

Critical Thinking—*Drawing Conclusions*
5. Do you think there is a need for catalysts that slow down a reaction? Give some examples.

CONNECTIONS

Chemical Reactions That Destroy the Environment

Stop and look around for a moment. Do you notice that you are surrounded by all sorts of devices that make life easier and more comfortable? There are cars, airplanes, air conditioners, televisions, refrigerators, radios, and industrial machinery—to mention a few. Where do all these inventions get their energy? Most of it comes from the burning of coal and petroleum.

Coal and petroleum are fuels that contain the element carbon. When they burn (a combustion reaction), carbon dioxide (CO_2) is released. The combustion of coal and petroleum releases about 20 billion tons of carbon dioxide into the atmosphere every year. And carbon dioxide in the atmosphere is of great consequence. The Earth and its inhabitants send a great deal of heat out through the atmosphere toward space. It is vital that this heat escapes the Earth's atmosphere. If it does not, the surface of the Earth will heat up, changing climates significantly. Although carbon dioxide is only a small part of the atmosphere, it has the ability to trap heat near the surface of the Earth. By absorbing heat released from the Earth, carbon dioxide acts like the glass over a greenhouse. For this reason, the warming of the Earth by carbon dioxide and other heat-trapping gases is referred to as the *greenhouse effect*. Some climatologists (scientists who study climate) believe that the buildup of carbon dioxide and other gases has already begun to change the climate of our planet. They cite as evidence the fact that the six warmest years on record were all in the 1980s. The trend is continuing into the 1990s.

One long-term outcome of the warming of the Earth (global warming) is the melting of the polar icecaps. As the icecaps melt because of higher temperatures, the sea level rises, causing certain coastal areas to be flooded and destroyed. Another effect is a shift in the distribution of rainfall over the various continents. Eventually, heat and drought could become commonplace over the vast region of the United States that now provides most of the world's agricultural products.

In order to minimize and possibly reverse the greenhouse effect, new energy sources that do not release carbon dioxide into the atmosphere must be developed. Energy must also be conserved. What role do you think you can play in making sure chemical reactions do not further destroy the environment?

Laboratory Investigation

Determining Reaction Rate

Problem

How does concentration affect reaction rate?

Materials (per group)

safety goggles
2 graduated cylinders
120 mL Solution A
3 250-mL beakers
distilled water at room temperature
90 mL Solution B
stirring rod
sheet of white paper
stopwatch or watch with a sweep second
    hand

Procedure 🧪 📦 👁

1. Carefully measure 60 mL of Solution A and pour it into a 250-mL beaker. Add 10 mL of distilled water and stir.

2. Carefully measure 30 mL of Solution B and pour it into a second beaker. Place the beaker of Solution B on a sheet of white paper in order to see the color change more easily.

3. Add the 70 mL of Solution A-water mixture to Solution B. Stir rapidly. Record the time it takes for the reaction to occur.

4. Rinse and dry the reaction beaker.

5. Repeat the procedure using the other amounts shown in the data table.

Observations

Solution A (mL)	Distilled Water Added to Solution A (mL)	Solution B (mL)	Reaction Time (sec)
60	10	30	
40	30	30	
20	50	30	

1. What visible indication is there that a chemical reaction is occurring?

2. What is the effect of adding more distilled water on the concentration of Solution A?

3. What happens to reaction time as more distilled water is added to Solution A?

4. Make a graph of your observations by plotting time along the X axis and volume of Solution A along the Y axis.

Analysis and Conclusions

1. How does concentration affect reaction rate?

2. Does your graph support your answer to question 1? Explain why.

3. What would a graph look like if time were plotted along the X axis and volume of distilled water added to Solution A were plotted along the Y axis?

4. In this investigation, what is the variable? The control?

5. **On Your Own** Enhance your graph by testing other concentrations.

Summarizing Key Concepts

2–1 Nature of Chemical Reactions

▲ When a chemical reaction occurs, there is always a change in the properties and the energy of the substances.

▲ A reactant is a substance that enters into a chemical reaction. A product is a substance that is produced by a chemical reaction.

2–2 Chemical Equations

▲ The law of conservation of mass states that matter can be neither created nor destroyed in a chemical reaction.

▲ A chemical equation that has the same number of atoms of each element on both sides of the arrow is a balanced equation.

2–3 Types of Chemical Reactions

▲ In a synthesis reaction, two or more simple substances combine to form a new, more complex substance.

▲ In a decomposition reaction, a complex substance breaks down into two or more simpler substances.

▲ In a single-replacement reaction, an un-combined element replaces an element that is part of a compound.

▲ In a double-replacement reaction, different atoms in two different compounds replace each other.

2–4 Energy of Chemical Reactions

▲ Energy is released in an exothermic reaction. Energy is absorbed in an endothermic reaction.

▲ In order for reactants to form products, activation energy is needed.

2–5 Rates of Chemical Reactions

▲ The rate of a reaction is a measure of how quickly reactants turn into products.

▲ An increase in the concentration of reactants increases the rate of a reaction.

▲ An increase in the surface area of reactants increases the rate of reaction.

▲ An increase in temperature generally increases the rate of a reaction.

Reviewing Key Terms

Define each term in a complete sentence.

2–1 Nature of Chemical Reactions
chemical reaction
reactant
product

2–2 Chemical Equations
chemical equation

2–3 Types of Chemical Reactions
synthesis reaction
decomposition reaction

single-replacement reaction
double-replacement reaction

2–4 Energy of Chemical Reactions
exothermic reaction
endothermic reaction
activation energy

2–5 Rates of Chemical Reactions
kinetics
reaction rate
collision theory
catalyst

Chapter Review

Content Review

Multiple Choice

Choose the letter of the answer that best completes each statement.

1. In a balanced chemical equation,
   a. atoms are conserved.
   b. molecules are equal.
   c. coefficients are equal.
   d. energy is not conserved.
2. Two or more substances combine to form one substance in a
   a. decomposition reaction.
   b. double-replacement reaction.
   c. single-replacement reaction.
   d. synthesis reaction.
3. In an endothermic reaction, heat is
   a. absorbed.   c. destroyed.
   b. released.   d. conserved.
4. The energy required for reactants to form products is called
   a. energy of motion.
   b. potential energy.
   c. activation energy.
   d. synthetic energy.

5. The substances to the left of the arrow in a chemical equation are called
   a. coefficients.   c. subscripts.
   b. products.   d. reactants.
6. An atom's ability to undergo chemical reactions is determined by
   a. protons.   c. innermost electrons.
   b. neutrons.   d. outermost electrons.
7. The rate of a chemical reaction can be increased by
   a. decreasing concentration.
   b. increasing surface area.
   c. removing a catalyst.
   d. all of these.
8. Adding a catalyst to a reaction increases its rate by
   a. increasing molecular motion.
   b. decreasing molecular motion.
   c. lowering activation energy.
   d. increasing concentration.

True or False

If the statement is true, write "true." If it is false, change the underlined word or words to make the statement true.

1. The substances formed as a result of a chemical reaction are called <u>reactants</u>.
2. A number written in front of a chemical symbol or formula is a(an) <u>coefficient</u>.
3. In a <u>synthesis</u> reaction, complex substances form simpler substances.
4. The formation of carbon dioxide during combustion of a fuel is an example of a <u>decomposition</u> reaction.
5. In an exothermic reaction, products have <u>more</u> energy than reactants.
6. The study of the rates of chemical reactions is <u>kinetics</u>.
7. The <u>collision theory</u> can be used to account for the factors that affect reaction rates.

Concept Mapping

Complete the following concept map for Section 2–2. Refer to pages O6–O7 to construct a concept map for the entire chapter.

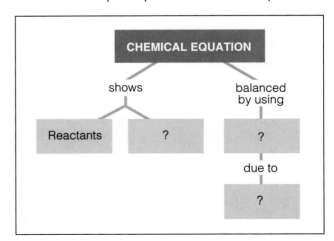

Concept Mastery

Discuss each of the following in a brief paragraph.

1. What is a chemical reaction? Why do substances react chemically?
2. Describe what a chemical equation is.
3. State the law of conservation of mass and explain its role in chemical equations.
4. What is a coefficient? What is the difference between H_2O_2 and $2H_2O$?
5. Describe each of the four types of chemical reactions. Include an equation showing the general form for each.
6. What is activation energy? How is it affected by a catalyst?
7. Explain how the energy content of the products of a reaction compares with that of the reactants when the reaction is exothermic. When it is endothermic.
8. Use the collision theory to explain the effects on reaction rate of increased concentration, catalysts, and increased surface area.
9. Give two reasons why collisions between molecules of reactants may not be effective in forming products.

Critical Thinking and Problem Solving

Use the skills you have developed in this chapter to answer each of the following.

1. **Making calculations** Balance the following equations:
   a. $PbO_2 \longrightarrow PbO + O_2$
   b. $Ca + H_2O \longrightarrow Ca(OH)_2 + H_2$
   c. $Zn + S \longrightarrow ZnS$
   d. $BaCl_2 + Na_2SO_4 \longrightarrow BaSO_4 + NaCl$
   e. $Al + Fe_2O_3 \longrightarrow Al_2O_3 + Fe$
   f. $C_{12}H_{22}O_{11} \longrightarrow C + H_2O$
2. **Classifying reactions** Identify the general type of reaction represented by each equation. Explain your answers.
   a. $Fe + 2HCl \longrightarrow FeCl_2 + H_2$
   b. $NiCl_2 \longrightarrow Ni + Cl_2$
   c. $4C + 6H_2 + O_2 \longrightarrow 2C_2H_6O$
   d. $2LiBr + Pb(NO_3)_2 \longrightarrow 2LiNO_3 + PbBr_2$
   e. $CaO + H_2O \longrightarrow Ca(OH)_2$
3. **Developing a model** Draw an energy diagram of an exothermic reaction that has a high activation energy. On your diagram, indicate how an increase in temperature would affect the rate of this reaction. Do the same for the addition of a catalyst.
4. **Applying concepts** Why can a lump of sugar be used in hot tea but granulated (small crystals) sugar is preferred in iced tea?

5. **Recognizing relationships** Potential energy is energy that is stored for later use. Explain where the potential energy is located in an exothermic reaction? Why do foods and fuel have chemical potential energy?
6. **Using the writing process** You are hired by a large company to design an advertising campaign. The company galvanizes, or covers, iron with the more active metal zinc to protect iron from corroding. Develop an advertising slogan and poster that explain to the public why this process is important.

Families of Chemical Compounds

Guide for Reading

After you read the following sections, you will be able to

3–1 Solution Chemistry
- Define a solution and describe its properties.

3–2 Acids and Bases
- Describe properties and uses of acids and bases.

3–3 Acids and Bases in Solution: Salts
- Relate pH number to acid-base strength.
- Describe the formation of salts.

3–4 Carbon and Its Compounds
- Describe organic compounds.

3–5 Hydrocarbons
- Identify three series of hydrocarbons.

3–6 Substituted Hydrocarbons
- Explain the nature of substituted hydrocarbons.

Banana, strawberry, pineapple, peach—what's your special flavor? How would you order your favorite ice cream sundae or soda? Certainly not by asking for a scoop of methyl butylacetate! Or a solution of carbonic acid, lactic acid, and ethyl butyrate! Yet that is exactly what you are eating when you enjoy a banana ice cream sundae or a pineapple ice cream soda.

Methyl butylacetate is the banana-flavored compound that makes banana ice cream different from strawberry or vanilla ice cream. And pineapple ice cream owes its characteristic flavor to the presence of ethyl butyrate.

Methyl butylacetate and ethyl butyrate belong to a much larger group of compounds known as organic compounds. Organic compounds are also found in the sugar and cream in the ice cream. But that's not all. In addition to organic compounds, your ice cream sundae or soda contains compounds known as acids, bases, and salts.

In this chapter you will learn about several groups of important compounds—compounds with a variety of surprising and sometimes delicious uses. So sit back and think about a nice big ethyl cinnemate sundae topped with isoamyl salicylate. . . .

Journal *Activity*

You and Your World What do you think of when you hear the words solution, saturated, concentration, base, salt, and neutral? In your journal, write your immediate reaction to each word. When you have finished reading this chapter, see how closely your definitions match the scientific ones.

◀ *What delicious examples of the many uses of two important families of chemical compounds!*

3–1 Solution Chemistry

In Chapter 1 you learned that as a result of chemical bonding (the combining of atoms of elements to form new substances), hundreds of thousands of different substances exist. In Chapter 2 you were introduced to the various chemical reactions by which atoms are rearranged to form new and different substances. In this chapter you will discover how scientists have attempted to bring order to the incredible number of different types of chemical compounds that exist. In other words, you will learn about a system of classifying compounds into families based on their physical and chemical properties.

One of the most important and abundant families of chemical compounds is the family of acids, bases, and salts. This family may already be familiar to you. Have you ever heard of acetic acid in vinegar? Magnesium hydroxide in milk of magnesia? Or sodium hydrogen carbonate in baking soda? Acetic acid is an acid; magnesium hydroxide is a base; and sodium hydrogen carbonate is a salt. See—you do know something about this family!

In order to better understand acids, bases, and salts, it will be helpful to take a few steps back and look at an important process that produces these compounds. The process involves making solutions.

Figure 3–1 *The solution process has produced some of nature's loveliest wonders. These stalactites and stalagmites, found in Luray Caverns, Virginia, formed when salts crystallized out of solution. Acids, bases, and salts are found in many common substances, such as these appetizing breads.*

What Is a Solution?

What happens when a lump of sugar is dropped into a glass of lemonade? What takes place when carbon dioxide gas is bubbled through water? And where do mothballs go when they disappear? The answer to these questions is the same: The sugar, gaseous carbon dioxide, and mothballs all dissolve in the substances in which they are mixed.

Careful examination of each of these mixtures—even under a microscope—will not reveal molecules of sugar in lemonade, carbon dioxide in water, or naphthalene in the air. But the sweet taste of lemonade tells you the sugar is there. The "fizziness" of soda water indicates the presence of carbon dioxide. And the smell of mothballs is evidence of naphthalene. In each of these mixtures, the molecules of one substance have become evenly distributed among the molecules of the other substance. The mixtures are uniform (the same) throughout. Each mixture is a **solution.**

Figure 3–2 *Solutions abound in nature. Blood is a solution. It contains fats, salts, sugars, and proteins dissolved in water. The dragon that guards the Forbidden City in China is made of bronze, which is a type of solution known as an alloy. A solution of limestone rock and water formed a cavern, which then collapsed and formed a sinkhole . What solutions can you identify in this scene of Glacier Park, Montana?*

ACTIVITY DOING

Emulsifying Action

1. To 10 mL of oil, add an equal amount of vinegar. Do the two liquids dissolve in each other?

2. Shake the liquids and observe what immediately happens. What happens after several minutes?

3. Now add an egg yolk to the mixture and shake again. Describe what happens.

The egg yolk acts as an emulsifying agent, changing one type of mixture to another.

4. Find out what the words colloid and suspension mean.

In this activity, which is the colloid? The suspension?

What is a common name for your end product?

A solution is a mixture in which one substance is dissolved in another substance. Different parts of a solution are identical. That is what is meant by the words uniform throughout. The molecules making up a solution are too small to be seen and do not settle when the solution is allowed to stand. A solution, then, is a "well-mixed" mixture.

All solutions have several basic properties. Let's go back to the glass of sweetened lemonade to discover just what these properties are. **A solution consists of two parts: One part is the substance being dissolved, and the other part is the substance doing the dissolving.**

The substance that is dissolved is called the **solute** (SAHL-yoot). The substance that does the dissolving is called the **solvent** (SAHL-vuhnt). The solvent is sometimes called the dissolving medium. In the sweetened lemonade, the solute is the sugar and the solvent is the lemonade. Even without the sugar, the lemonade is a solution. It is made of water and lemon juice. What are the solutes and solvents in the other two examples of solutions you have just read about?

The most common solutions are those in which the solvent is a liquid. The solute can be a solid, a liquid, or a gas. The most common solvent is water.

Figure 3–3 *A solution consists of a solute and a solvent. The most common solutions are those in which the solvent is a liquid. Here you see the three types of solutions that can be formed from a liquid solvent and a solid, a liquid, and a gas solute. What is the most common liquid solvent?*

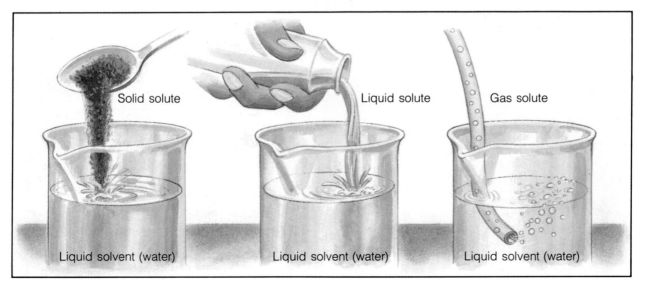

Solid solute

Liquid solute

Gas solute

Liquid solvent (water)

Liquid solvent (water)

Liquid solvent (water)

Figure 3–4 *A liquid solution appears clear. Rocks known as geodes form when a solvent evaporates, leaving behind deposits of solute.*

When alcohol is the solvent, the solution is called a tincture. Have you ever heard of tincture of iodine? It is an antiseptic used to treat minor cuts and scratches. What is the solute in this solution?

The particles in a solution are individual atoms, ions, or molecules. Because the particles are so small, they do not scatter light that passes through the solution. A liquid solution appears clear.

Most solutions cannot easily be separated by simple physical means such as filtering. However, a physical change such as evaporation or boiling can separate the parts of many solutions. If salt water is boiled, the water will change from a liquid to a gas, leaving behind particles of salt.

Another important property of a solution is its ability or inability to conduct an electric current. An electric current is a flow of electrons. In order for electrons to flow through a solution, ions must be present. As you learned in Chapter 1, ions are charged atoms. A solution that contains ions is a good conductor of electricity. A solution that does not contain ions is a nonconductor.

Pure water is a poor conductor of electric current because it does not contain ions. If a solute such as potassium chloride (KCl) is added to the water, however, the resulting solution is a good conductor. Substances whose water solutions conduct an electric current are called **electrolytes** (ee-LEHK-troh-lights). Potassium chloride, sodium chloride, and silver nitrate are examples of electrolytes. Most electrolytes are ionic compounds.

Substances whose water solutions do not conduct an electric current are called **nonelectrolytes.** A solution of sugar and water does not conduct an electric current. Sugar is a nonelectrolyte, as are alcohol and benzene. Many covalent compounds are nonelectrolytes because they do not form ions in solution.

Making Solutions

Solutions abound in nature. The oceans, the atmosphere, even Earth's interior are solutions. Your body contains a number of vital solutions. Every solution has a particular solute and solvent.

TYPES OF SOLUTIONS

Solute	Solvent	Example
Gas Gas Gas	Gas Liquid Solid	Air (oxygen in nitrogen) Soda water (carbon dioxide in water) Charcoal gas mask (poisonous gases on carbon)
Liquid Liquid Liquid	Gas Liquid Solid	Humid air (water in air) Antifreeze (ethylene glycol in water) Dental filling (mercury in silver)
Solid Solid Solid	Gas Liquid Solid	Soot in air (carbon in air) Ocean water (salt in water) Gold jewelry (copper in gold)

Figure 3–5 *Nine different types of solutions can be made from the three phases of matter. What are solutions of solids dissolved in solids called?*

TYPES OF SOLUTIONS   Matter can exist as a solid, a liquid, or a gas. From these three phases of matter, nine different types of solutions can be made. Figure 3–5 shows these types of solutions.

The most common solutions are liquid solutions, or solutions in which the solvent is a liquid. The solute can be a solid, a liquid, or a gas. Two liquids that dissolve in each other are said to be miscible (MIHS-uh-buhl). Water and alcohol are miscible. Do you think oil and water are miscible? Solutions of solids dissolved in solids are called alloys. Most alloys are made of metals. Some common alloys include brass, bronze, solder, stainless steel, and wrought iron. You might want to find out exactly what the solute and solvent are in each of these alloys.

RATE OF SOLUTION   Suppose you wanted to dissolve some sugar in a glass of water as quickly as possible. What might you do? If your answer included stirring the solution, using granulated sugar, or heating the water, you are on the right track.

Normally, the movement of solute molecules away from the solid solute and throughout the solvent occurs rather slowly. Stirring or shaking the solution helps to move solute particles away from the solid solute faster. More molecules of the solute are brought in contact with the solvent sooner, so the solute dissolves at a faster rate.

Figure 3–6 *Scientists have developed a model to describe a probable solution process. First, solute particles separate from the surface of the solid solute. Then, the solute molecules enter the liquid surrounding the solid solute. Finally, the solute molecules are attracted to the solvent molecules.*

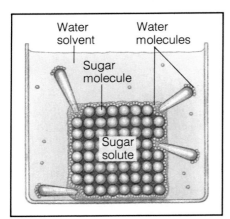

Solution action occurs only at the surface of the solid solute. So if the surface area of the solute is increased, the rate of solution is increased. When the solid solute is ground into a fine powder, more solute molecules are in contact with the solvent. Finely powdered solids dissolve much faster than do large lumps or crystals of the same substance.

If heat is applied to a solution, the molecules move faster and farther apart. As a result, the dissolving action is speeded up.

SOLUBILITY From experience, you know that table salt and sugar readily dissolve in water. These compounds are described as being very soluble or having a high degree of **solubility** (sahl-yoo-BIHL-uh-tee) in water. The solubility of a solute is a measure of how much of that solute can be dissolved in a given amount of solvent under certain conditions. Although a large amount of table salt dissolves in water, only a small amount of table salt dissolves in alcohol. So although the solubility of salt in water is high, the solubility of salt in alcohol is rather low. The solubility of a solute depends on the nature of the solute and of the solvent. Solubility is usually described in terms of the mass of the solute that can be dissolved in a definite amount of solvent at a specific temperature.

Two main factors affect the solubility of a solute. These factors are temperature and pressure. Usually, an increase in the temperature of a solution increases the solubility of a solid in a liquid. Just think how much more sugar dissolves in hot tea than in cold.

ACTIVITY
DISCOVERING

Solubility of a Gas in a Liquid

1. Remove the cap from a bottle of soda.

2. Immediately fit the opening of a balloon over the top of the bottle. Shake the bottle several times. Note any changes in the balloon.

3. Heat the bottle of soda very gently by placing it in a pan of hot water. Note any further changes in the balloon.

■ What two conditions of solubility are being tested?

■ What general statement about the solubility of a gas in a liquid can you now make?

Figure 3–7 *The solubility of a gas solute in a liquid solvent depends on both the pressure and the temperature. When the cap is removed from the bottle, the solubility of the gas decreases. If the bottle is cold, the decrease in solubility is very small. If the bottle is warm, the decrease is obvious.*

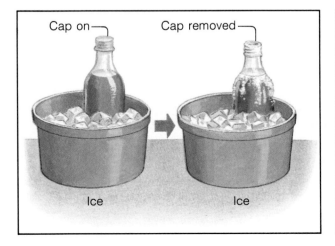

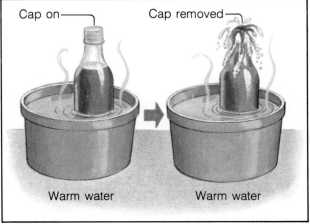

Figure 3–8 *This sight may be a common one to you — and now you know enough about solubility to explain it. What is happening here?*

ACTIVITY

DISCOVERING

Sudsing Power

1. Separately add 0 mL, 10 mL, 20mL, and 40 mL of salt to four 1-liter jars each containing 500 mL of water. Stir until the salt dissolves.

2. Add 10 mL of laundry soap to each jar.

3. Cap each jar and vigorously shake for 20 seconds.

■ How did the different concentrations of salt affect the sudsing action of the laundry soap?

■ What effect does water that contains large amounts of dissolved minerals have on washing clothes and on bathing? Do you know the name of this type of water?

The situation for a gas dissolved in a liquid is just the opposite. Raising the temperature of a solution decreases the solubility of a gas in a liquid. Perhaps you have observed this fact without actually realizing it. Have you ever let a glass of soda get warm? If so, what did you notice? The soda goes flat, or loses its fizz. The fizz in soda is carbon dioxide gas dissolved in soda water. As the temperature of the solution increases, the solubility of the carbon dioxide decreases. The gas comes out of solution, leaving the soda flat. Why do you think boiled water tastes flat?

For both solids and liquids dissolved in liquids, increases and decreases in pressure have practically no effect on solubility. For gases dissolved in liquids, however, an increase in pressure increases solubility, and a decrease in pressure decreases solubility. A bottle of soda fizzes when the cap is removed because molecules of carbon dioxide gas escape from solution as the pressure is decreased. The solubility of the carbon dioxide gas has been decreased by a decrease in pressure.

CONCENTRATION The **concentration** of a solution refers to the amount of solute dissolved in a certain amount of solvent. A solution in which a lot of solute is dissolved in a solvent is called a **concentrated solution.** A solution in which there is little solute dissolved in a solvent is called a **dilute solution.** The terms concentrated and dilute are not precise, however. They do not indicate exactly how much solute and solvent are present.

Using the concept of solubility, the concentration of a solution can be expressed in another way. A solution can be described as saturated, unsaturated, or supersaturated. To understand these descriptions, remember that solubility measures the maximum amount of solute that can be dissolved in a given amount of solvent.

A **saturated solution** is a solution that contains all the solute it can possibly hold at a given temperature. In a saturated solution, no more solute can be dissolved at that temperature. If more solute is added to a saturated solution, it will settle undissolved to the bottom of the solution. In describing a saturated solution, the temperature must always be given. Can you explain why?

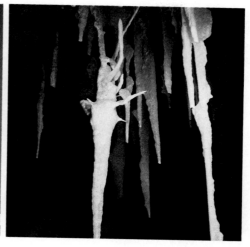

An **unsaturated solution** is a solution that contains less solute than it can possibly hold at a given temperature. In an unsaturated solution, more solute can be dissolved.

Under special conditions, a solution can be made to hold more solute than is normal for that temperature. Such a solution is a **supersaturated solution.** A supersaturated solution is unstable. If a single crystal of solute is added to a supersaturated solution, the excess solute comes out of solution and settles to the bottom. Only enough solute to make the solution saturated remains dissolved.

SPECIAL PROPERTIES Why is salt spread on roads and walkways that are icy? Why is salt added to cooking water? Why is a substance known as ethylene glycol added to the cooling systems of cars? The answers to these questions have to do with two special properties of solutions.

Experiments show that when a solute is dissolved in a liquid solvent, the freezing point of the solvent is lowered. The lowering of the freezing point is called freezing point depression. The addition of solute molecules interferes with the phase change of solvent molecules. So the solution can exist in the liquid phase at a lower temperature than can the pure solvent. Ethylene glycol, commonly known as antifreeze, is added to automobile cooling systems to lower the freezing point of water.

The addition of a solute to a pure liquid solvent also raises the boiling point of the solvent. This increase is called boiling point elevation. In this case,

Figure 3–9 *A saturated solution contains all the solute it can possibly hold at a given temperature. So any additional solute will fall to the bottom of the solution (left). Crystals of sugar are growing on a string placed in a supersaturated solution of sugar and water (center). Adding a small crystal of solute starts the crystallizing action instantly. The formation of these stalactites is also the result of solute crystallization from a supersaturated solution (right).*

Figure 3–10 *Salt is spread on icy surfaces because it causes the ice to melt at a lower temperature. Rock salt is used in an ice cream maker to lower the freezing point of the ice that surrounds the container of ice cream. This means that the ice melts at and remains at a temperature below 0°C. This, in turn, allows for more efficient cooling of the ice cream mixture and the eventual freezing of the ice cream.*

the addition of solute molecules interferes with the rapid evaporation, or boiling, of the solvent molecules. So the solution can exist in the liquid phase at a higher temperature than can the pure solvent. When salt is added to cooking water, the water boils at a higher temperature. Although it takes a longer time to heat the water to boiling, it takes a shorter time to cook food in that water.

3–1 Section Review

1. What is one important family of compounds? What process produces these compounds?
2. Compare a solute and a solvent.
3. What are three properties of a solution?
4. Describe three ways in which you can increase the rate at which a solid dissolves in a liquid.
5. Compare a saturated, unsaturated, and supersaturated solution.

Connection—*You and Your World*
6. Why is the antifreeze ethylene gylcol used in automobile radiators in warm weather?

3–2 Acids and Bases

If you look in your medicine cabinet and refrigerator and on your kitchen shelves, you will find examples of groups of compounds known as **acids** and **bases.** Acids are found in aspirin, vitamin C, and eyewash. Fruits such as oranges, grapes, lemons, grapefruits, and apples contain acids. Milk and tea contain acids, as do pickles, vinegar, and carbonated drinks. Bases are found in products such as lye, milk of magnesia, deodorants, ammonia, and soaps.

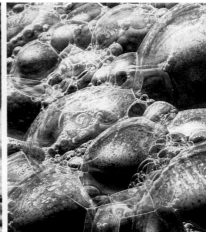

Figure 3–11 *Many of the fruits you eat contain acids. Acids are also used in the manufacture of synthetic fibers. The soap that produced these bubbles was manufactured using bases.*

Acids and bases also play an important role in the life processes that take place in your body. Many industrial processes use acids and bases. The manufacture of a wide variety of products involves the use of acids and bases.

Properties of Acids

As a class of compounds, all acids have certain physical and chemical properties when dissolved in water. One of the physical properties all acids share is sour taste. Lemons taste sour because they contain citric acid. Vinegar contains acetic acid. However, you should never use taste to identify a chemical substance. You should use other, safer properties.

Acids affect the color of indicators. Indicators are compounds that show a definite color change when mixed with an acid or a base. Litmus paper, a common indicator, changes from blue to red in an acid solution. Another indicator, phenolphthalein (fee-nohl-THAL-een), is colorless in an acid solution.

Acids react with active metals to form hydrogen gas and a metal compound. This reaction wears away, or corrodes, the metal and produces a residue. For example, sulfuric acid in a car battery often corrodes the terminals and leaves a residue.

Another property of acids can be identified by looking at the list of common acids in Figure 3–12 on page 74. What do all these acids have in common? Acids contain hydrogen. When dissolved

Acid-Base Testing

1. Ask your teacher for several strips of red litmus paper and blue litmus paper.

2. Test these substances with the litmus paper to determine if they are an acid or a base: orange juice, milk, tea, coffee, soda, vinegar, ammonia cleaner, milk of magnesia, saliva.

Report the results of your tests to your classmates by preparing a poster in which you classify each substance.

in water, acids ionize to produce positive hydrogen ions (H^+). A hydrogen ion is a proton. So acids are often defined as proton donors.

The hydrogen ion, or proton, produced by an acid is quickly surrounded by a water molecule. The attraction between the hydrogen ion (H^+) and the water molecule (H_2O) results in the formation of a hydronium ion, H_3O^+.

The definition of an acid as a proton donor helps to explain why all hydrogen-containing compounds are not acids. Table sugar contains 22 hydrogen atoms, but it is not an acid. When dissolved in water, table sugar does not produce H^+ ions. Table sugar is not a proton donor. So it does not turn litmus paper red or phenolphthalein colorless.

Common Acids

The three most common acids in industry and in the laboratory are sulfuric acid (H_2SO_4), nitric acid (HNO_3), and hydrochloric acid (HCl). These three acids are strong acids. That means they ionize to a high degree in water and produce hydrogen ions. The presence of hydrogen ions makes strong acids good electrolytes. (Remember, electrolytes are good conductors of electricity.)

Acetic acid ($HC_2H_3O_2$), carbonic acid (H_2CO_3), and boric acid (H_3BO_3) are weak acids. They do not ionize to a high degree in water, so they produce few hydrogen ions. Weak acids are poor electrolytes. Figure 3–12 lists the name, formula, and uses of some common acids. Remember, handle any acid—weak or strong—with care!

Properties of Bases

When dissolved in water, all bases share certain physical and chemical properties. Bases usually taste bitter and are slippery to the touch. However, bases can be poisonous and corrosive. So you should never use taste or touch to identify bases.

COMMON ACIDS	
Name and Formula	**Uses**
Strong	
Hydrochloric HCl	Pickling steel; cleaning bricks and metals; digesting food
Sulfuric H_2SO_4	Manufacturing paints, plastics, fertilizers; dehydrating agent
Nitric HNO_3	Removing tarnish; making explosives (TNT); making fertilizers
Weak	
Carbonic H_2CO_3	Carbonating beverages
Boric H_3BO_3	Washing eyes
Phosphoric H_3PO_4	Making fertilizers and detergents
Acetic $HC_2H_3O_2$	Making cellulose acetate used in fibers and films
Citric $H_3C_6H_5O_7$	Making soft drinks

Figure 3–12 *The name, formula, and uses of some common acids are given in this chart. What is the difference between a strong acid and a weak acid?*

Figure 3–13 *The name, formula, and uses of some common bases are given in this chart. What ion do all these bases contain?*

Bases turn litmus paper from red to blue and phenolphthalein to bright pink. Bases emulsify, or dissolve, fats and oils. They do this by reacting with the fat or oil to form a soap. The base ammonium hydroxide is used as a household cleaner because it "cuts" grease. The strong base sodium hydroxide, or lye, is used to clean clogged drains.

All bases contain the hydroxide ion, OH^-. When dissolved in water, bases produce this ion. Because a hydroxide ion (OH^-) can combine with a hydrogen ion (H^+) and form water, a base is often defined as a proton (H^+) acceptor.

Common Bases

Strong bases dissolve readily in water to produce large numbers of ions. So strong bases are good electrolytes. Examples of strong bases include potassium hydroxide (KOH), sodium hydroxide ($NaOH$), and calcium hydroxide ($Ca(OH)_2$).

Weak bases do not produce large numbers of ions when dissolved in water. So weak bases are poor electrolytes. Ammonium hydroxide (NH_4OH) and aluminum hydroxide ($Al(OH)_3$) are weak bases. See Figure 3–13.

COMMON BASES	
Name and Formula	**Uses**
Strong	
Sodium hydroxide $NaOH$	Making soap; drain cleaner
Potassium hydroxide KOH	Making soft soap; battery electrolyte
Calcium hydroxide $Ca(OH)_2$	Leather production; making plaster
Magnesium hydroxide $Mg(OH)_2$	Laxative; antacid
Weak	
Ammonium hydroxide NH_4OH	Household cleaner
Aluminum hydroxide $Al(OH)_3$	Antacid; deodorant

3–2 Section Review

1. What are three important properties of acids? Of bases?
2. Why are acids called proton donors?
3. Why are bases called proton acceptors?
4. If an electric conductivity setup were placed in the following solutions, would the light be bright or dim: HCl, HNO_3, H_3BO_3, $HC_2H_3O_2$, NH_4OH, KOH, $NaOH$, $Al(OH)_3$?

Connection—*Laboratory Safety*
5. How could you safely determine whether an unknown solution is an acid or a base?

3–3 Acids and Bases in Solution: Salts

As you have just learned, solutions can be acidic or basic. Solutions can also be neutral. To measure the acidity of a solution, the **pH** scale is used. **The pH of a solution is a measure of the hydronium ion (H_3O^+) concentration.** Remember, the hydronium ion is formed by the attraction between a hydrogen ion (H^+) from an acid and a water molecule (H_2O). So the pH of a solution indicates how acidic the solution is.

The pH scale is a series of numbers from 0 to 14. The middle of the scale—7—is the neutral point. A neutral solution has a pH of 7. It is neither an acid nor a base. Water is a neutral liquid.

A solution with a pH below 7 is an acid. Strong acids have low pH numbers. Would hydrochloric acid have a pH closer to 2 or to 6?

A solution with a pH above 7 is a base. Strong bases have high pH numbers. What would be the pH of NaOH?

Figure 3–14 *Does it surprise you to learn that many of the substances you use every day contain acids and bases? Which fruit is most acidic? What cleaner is most basic?*

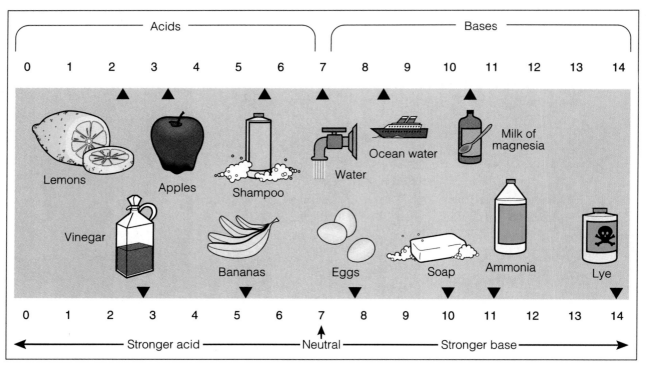

Determining Solution pH

The pH of a solution can be determined by using an indicator. You already know about two indicators: litmus paper and phenolphthalein. Other indictors include pH paper, methyl orange, and bromthymol blue. Each indicator shows a specific color change as the pH of a solution changes.

Common household materials can be used as indicators. Red-cabbage juice covers the entire pH range. Grape juice is bright pink in the presence of an acid and bright yellow in the presence of a base. Even tea can be an indicator. Have you ever noticed the color of tea change when you add lemon juice? For accurate pH measurements, a pH meter is used.

Formation of Salts

When acids react chemically with bases, they form a class of compounds called salts. A **salt** is a compound formed from the positive ion of a base and the negative ion of an acid. A salt is a neutral substance.

The reaction of an acid with a base produces a salt and water. The reaction is called **neutralization** (noo-truhl-ih-ZAY-shuhn). In neutralization, the properties of the acid and the base are lost as two neutral substances—a salt and water—are formed.

The reaction of HCl with NaOH is a neutralization reaction. The positive hydrogen ion from the acid combines with the negative hydroxide ion from the base. This produces water. The remaining positive ion of the base combines with the remaining negative ion of the acid to form a salt.

$$H^+Cl^- + Na^+OH^- \longrightarrow H_2O + NaCl$$

Many of the salts formed by a neutralization reaction are insoluble in water—that is, they do not dissolve in water. They crystallize out of solution and remain in the solid phase. An insoluble substance that crystallizes out of solution is called a precipitate (pree-SIHP-uh-tayt). The process by which a precipitate forms is called precipitation. An example of a precipitate is magnesium carbonate. Snow, rain, sleet, and hail are considered forms of precipitation because they fall out of solution. Out of what solution do they precipitate?

ACTIVITY DOING

A Homemade Indicator

1. Shred two leaves of red cabbage and boil them in some water until the liquid gets very dark. **CAUTION:** *Wear safety goggles and use extreme care .*

2. When the liquid and the shreds have cooled, squeeze all the purple juice you can from the shreds. Pour the liquid into a container.

3. Into 5 separate containers, pour about 3 to 4 mL of the following substances, one in each container: shampoo, grapefruit juice, clear soft drink, milk, ammonia cleaner.

4. Add about 2 mL of the cabbage liquid. Stir and observe any color changes.

Based on the fact that ammonia is a base, identify the other substances as either acids or bases.

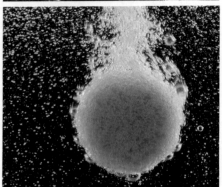

Figure 3–15 *To control dangerous acid spills, firefighters spray a blanket of base over the acid. The result is a neutral salt and water. These two substances are also produced when a person uses an antacid to relieve an upset stomach. What reaction is taking place in both of these situations?*

A neutralization reaction is a double-replacement reaction—and a very important one, too! For a dangerous acid can be combined with a dangerous base to form a harmless salt and neutral water.

3–3 Section Review

1. What is pH? Describe the pH scale.
2. What is the pH of an acid? A base? Water?
3. How can the pH of a solution be determined?
4. How does a salt form? What is a salt?
5. What is neutralization?

Critical Thinking—*Applying Concepts*
6. Use an equation to show the neutralization reaction between H_2SO_4 and NaOH.

Guide for Reading

Focus on these questions as you read.
▶ *What are organic compounds?*
▶ *What are some properties of organic compounds?*

3–4 Carbon and Its Compounds

Do you know what such familiar substances as sugar, plastic, paper, and gasoline have in common? They all contain the element carbon. Carbon is present in more than 2 million known compounds, and this number is rapidly increasing. Approximately 100,000 new carbon compounds are being isolated or synthesized every year. In fact, more than 90 percent of all known compounds contain carbon! **Carbon compounds form an important family of chemical compounds known as organic compounds.** The word organic means coming from life. Because carbon-containing compounds are present in all living things, scientists once believed that **organic compounds** could be produced only by living organisms. Living things were thought to have a mysterious "vital force" that was responsible for creating

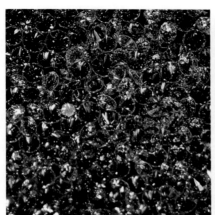
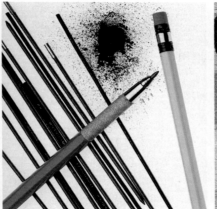

Figure 3–16 *The element carbon is present in more than 2 million known compounds. Here you see three different forms of the pure element: diamond, graphite, and coal.*

carbon compounds. It was believed that the force could not be duplicated in the laboratory.

In 1828, the German chemist Friedrich Wöhler produced an organic compound called urea from two inorganic substances. Urea is a waste product produced by the human body. It was not long before chemists accepted the idea that organic compounds could be prepared from materials that were never part of a living organism. What is common to all organic compounds is not that they originated in living things but that they all contain the element carbon. Today, the majority of organic compounds are synthesized in laboratories.

There are some carbon compounds that are not considered organic compounds. Calcium carbonate, carbon dioxide, and carbon monoxide are considered inorganic (not organic) compounds.

The Bonding of Carbon

Carbon's ability to combine with itself and with other elements explains why there are so many carbon compounds. Carbon atoms form covalent bonds with other carbon atoms.

The simplest bond involves 2 carbon atoms. The most complex involves thousands of carbon atoms. The carbon atoms can form long straight chains, branched chains, single rings, or rings joined together.

The bonds between carbon atoms can be single covalent bonds, double covalent bonds, or triple covalent bonds. In a single bond, one pair of electrons is shared between 2 carbon atoms. In a double

ACTIVITY

DOING

Octane Rating

1. Find out what the octane rating of gasoline means.

2. Go to a local gas station and find out the octane ratings of the different grades of gasoline being sold. Compare the prices of the different grades. Ask the station attendant to describe how each grade of gasoline performs.

3. Present your findings to your class.

Figure 3–17 *Because of carbon's bonding ability, a great variety of organic compounds exists. These include synthetic rubber and plastic for firefighters' equipment, paints and dyes, and the substances of which all living things are made.*

bond, two pairs of electrons are shared between 2 carbon atoms. See Figure 3–18. How many pairs of electrons are shared in a triple bond?

Carbon atoms also bond with many other elements. These elements include oxygen, hydrogen, members of the nitrogen family, and members of Family 17. The simplest organic compounds contain just carbon and hydrogen. Because there are so many compounds of carbon and hydrogen, they form a class of organic compounds all their own. You will soon read about this class of compounds.

A great variety of organic compounds exists because the same atoms that bond together to form one compound may be arranged in several other ways in several other compounds. Each different arrangement of atoms represents a separate organic compound.

Properties of Organic Compounds

Organic compounds usually exist as gases, liquids, or low-melting-point solids. Organic liquids generally have strong odors and low boiling points. Organic liquids do not conduct an electric current. What is the name for a substance whose solution does not conduct electricity? Organic compounds generally do not dissolve in water. Oil, which is a mixture of organic compounds, floats on water because the two liquids are insoluble.

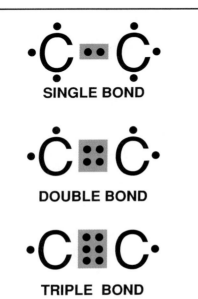

SINGLE BOND

DOUBLE BOND

TRIPLE BOND

Figure 3–18 *In a single bond, one pair of electrons is shared. In a double bond, two pairs of electrons are shared. How many pairs of electrons are shared in a triple bond?*

Structural Formulas

A molecular formula for a compound indicates what elements make up that compound and how many atoms of each element are present in a molecule. For example, the molecular formula for the organic compound ethane is C_2H_6. In every molecule of ethane, there are 2 carbon atoms and 6 hydrogen atoms.

What a molecular formula does not indicate about a molecule of a compound is how the different atoms are arranged. To do this, a **structural formula** is used. A structural formula shows the kind, number, and arrangement of atoms in a molecule. You can think of a structural formula as being a model of a molecule.

Figure 3–19 shows the structural formula for ethane and two other organic compounds: methane and propane. Note that in a structural formula, a dash (–) is used to represent the pair of shared electrons forming a covalent bond. In writing structural formulas, it is important that you remember the electron arrangement in a carbon atom.

Carbon has 4 valence electrons, or 4 electrons in its outermost energy level. Each electron will form a covalent bond with an electron of another atom to produce a stable outermost level containing 8 electrons. So when structural formulas are written, there can be no dangling bonds—no dangling dashes.

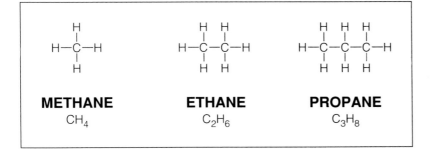

METHANE	ETHANE	PROPANE
CH_4	C_2H_6	C_3H_8

Figure 3–19 *Methane, ethane, and propane are the first three members of a series of hydrocarbons known as the alkanes. Note that each carbon atom is surrounded by four dashes, corresponding to four pairs of shared electrons.*

Isomers

Compounds with the same molecular formula but different structures are called **isomers**. Figure 3–20 on page 82 shows two isomers of butane, C_4H_{10}. Notice that one isomer is a straight chain and

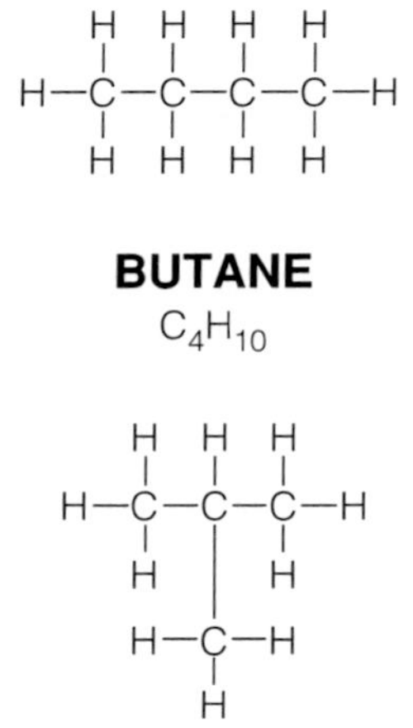

BUTANE
C_4H_{10}

ISOBUTANE
C_4H_{10}

Figure 3–20 *Butane has two isomers: normal butane and isobutane. Which isomer is a branched chain?*

the other isomer is a branched chain. In a branched chain, all the carbon atoms are not in a straight line. This difference in structure will account for any differences in the physical and chemical properties of these two compounds.

Figure 3–21 shows three isomers of pentane, C_5H_{12}. This time there is one straight chain and two branched chains. To see the difference between the two branched chains, count the number of carbon atoms in the straight-chain portion of each molecule. How many are there in each branched isomer? What do you think happens to the number of possible isomers as the number of carbon atoms in a molecule increases? The compound whose formula is $C_{15}H_{32}$ could have more than 400 isomers!

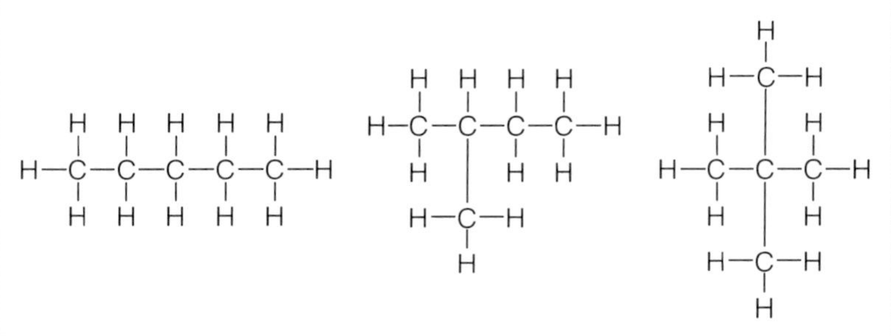

Figure 3–21 *As the number of carbon atoms increases, the number of isomers increases. What alkane is shown here? How do its three isomers differ?*

3–4 Section Review

1. What are organic compounds?
2. What four factors account for the abundance of carbon compounds?
3. What are three properties of organic compounds?

Critical Thinking—*Applying Concepts*
4. Could two compounds have the same structural formula but different molecular formulas?

3–5 Hydrocarbons

Have you ever heard of a butane lighter, seen a propane torch, or noticed a sign at a service station advertising "high octane" gasoline? Butane, propane, and octane are members of a large group of organic compounds known as **hydrocarbons.** A hydrocarbon contains only hydrogen and carbon.

Hydrocarbons can be classified as saturated or unsaturated depending on the type of bonds between carbon atoms. In **saturated hydrocarbons,** all the bonds between carbon atoms are single covalent bonds. In **unsaturated hydrocarbons,** one or more of the bonds between carbon atoms is a double covalent or triple covalent bond.

Alkanes

The **alkanes** are straight-chain or branched-chain hydrocarbons in which all the bonds between carbon atoms are single covalent bonds. Alkanes are saturated hydrocarbons. All the hydrocarbons that are alkanes belong to the alkane series. The simplest member of the alkane series is methane, CH_4. Methane consists of 1 carbon atom surrounded by 4 hydrogen atoms. Why are there 4 hydrogen atoms?

The next simplest alkane is ethane, C_2H_6. How does the formula for ethane differ from the formula for methane? After ethane, the next member of the

Figure 3–22 *Petroleum is one of the most abundant sources of hydrocarbons. The hydrocarbons in petroleum range from 1-carbon molecules to more than 50-carbon molecules. Petroleum, which does not mix with water, is highly flammable.*

Figure 3–23 *Common alkanes include methane, of which the planet Saturn's atmosphere is composed; propane, which is burned to provide heat for hot-air balloons; and butane, which is the fuel in most lighters. What is the general formula for the alkanes?*

ALKANE SERIES	
Name	**Formula**
Methane	CH_4
Ethane	C_2H_6
Propane	C_3H_8
Butane	C_4H_{10}
Pentane	C_5H_{12}
Hexane	C_6H_{14}
Heptane	C_7H_{16}
Octane	C_8H_{18}
Nonane	C_9H_{20}
Decane	$C_{10}H_{22}$

alkane series is propane, C_3H_8. Can you begin to see a pattern to the formulas for each successive alkane? Ethane has 1 more carbon atom and 2 more hydrogen atoms than methane does. Propane has 1 more carbon atom and 2 more hydrogen atoms than ethane does. Each member of the alkane series is formed by adding 1 carbon atom and 2 hydrogen atoms to the previous compound.

The pattern that exists for the alkanes can be used to determine the formula for any member of the series. Because each alkane differs from the preceding member of the series by the group CH_2, a general formula for the alkanes can be written. That general formula is C_nH_{2n+2}. The letter n is the number of carbon atoms in the alkane. What would be the formula for a 15-carbon hydrocarbon? For a 30-carbon hydrocarbon?

Naming Hydrocarbons

Figure 3–24 shows the first ten members of the alkane series. Look at the names of the compounds. How is each name the same? How is each different?

Often in organic chemistry the names of the compounds in the same series will have the same ending, or suffix. Thus, the members of the alkane series all end with the suffix *-ane*, the same ending as in the series name. The first part of each name, or the prefix, indicates the number of carbon atoms

Figure 3–24 *This chart shows the names and formulas for the first ten members of the alkane series. What does the prefix hex- mean?*

present in the compound. The prefix *meth-* indicates 1 carbon atom. The prefix *eth-*, 2 carbon atoms, and the prefix *prop-*, 3. According to Figure 3–24, how many carbon atoms are indicated by the prefix *pent-*? How many carbon atoms are in octane? As you study other hydrocarbon series, you will see that these prefixes are used again and again. So it will be useful for you to become familiar with the prefixes that mean 1 to 10 carbon atoms.

Alkenes

Hydrocarbons in which at least one pair of carbon atoms is joined by a double covalent bond are called **alkenes.** Alkenes are unsaturated hydrocarbons. The first member of the alkene series is ethene, C_2H_4. The next member of the alkene series is propene, C_3H_6.

Figure 3–25 shows the first seven members of the alkene series. What do you notice about the name of each compound?

As you look at the formulas for the alkenes, you will again see a pattern in the number of carbon and hydrogen atoms added to each successive compound. The pattern is the addition of 1 carbon atom and 2 hydrogen atoms. The general formula for the alkenes is C_nH_{2n}. The letter n is the number of carbon atoms in the compound. What is the formula for an alkene with 12 carbons? With 20 carbons?

In general, alkenes are more reactive than alkanes because a double bond is more easily broken than a single bond. So alkenes can react chemically by adding other atoms directly to their molecules.

Alkynes

Hydrocarbons in which at least one pair of carbon atoms is joined by a triple covalent bond are called **alkynes.** Alkynes are unsaturated hydrocarbons. The simplest alkyne is ethyne, C_2H_2, which is commonly known as acetylene. Perhaps you have heard of acetylene torches, which are used in welding.

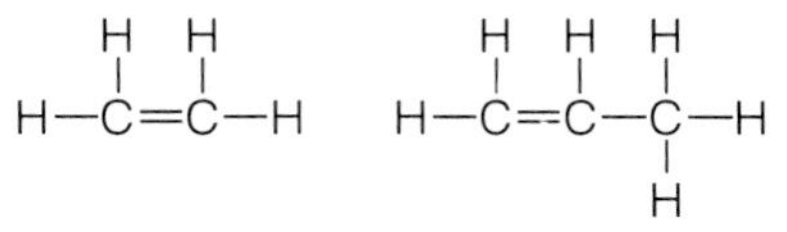

ALKENE SERIES	
Name	**Formula**
Ethene	C_2H_4
Propene	C_3H_6
Butene	C_4H_8
Pentene	C_5H_{10}
Hexene	C_6H_{12}
Heptene	C_7H_{14}
Octene	C_8H_{16}

Figure 3–25 *This chart shows the names and formulas of the first seven members of the alkene series. What would a 9-carbon alkene be called?*

ETHENE
C_2H_4

PROPENE
C_3H_6

Figure 3–26 *The first two members of the alkene series are ethene and propene. What kind of bonds do the alkenes have?*

ALKYNE SERIES

Name	Formula
Ethyne	C_2H_2
Propyne	C_3H_4
Butyne	C_4H_6
Pentyne	C_5H_8
Hexyne	C_6H_{10}

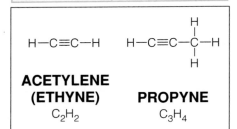

ACETYLENE
(ETHYNE)
C_2H_2

PROPYNE
C_3H_4

Figure 3–27 *This chart shows the names and formulas for the first five members of the alkyne series. What is the general formula for this series? The simplest alkynes are ethyne and propyne. What is the common name for ethyne?*

The first five members of the alkyne series are listed in Figure 3–27. Here again, each successive member of the alkyne series differs by the addition of 1 carbon atom and 2 hydrogen atoms. The general formula for the alkynes is C_nH_{2n-2}.

The alkynes are even more reactive than the alkenes. Very little energy is needed to break a triple bond. Like the alkenes, alkynes can react chemically by adding other atoms directly to their molecules.

Aromatic Hydrocarbons

All the hydrocarbons you have just learned about—the alkanes, alkenes, and alkynes—are either straight-chain or branched-chain molecules. But this is not the only structure a hydrocarbon can have. Some hydrocarbons are in the shape of rings. Probably the best-known class of hydrocarbons in the shape of rings is the aromatic hydrocarbons. The name of this class comes from the fact that aromatic hydrocarbons share a common physical property. These compounds have strong and often pleasant odors (or aromas).

Figure 3–28 *The simplest aromatic hydrocarbon is benzene. Benzene is used in the manufacture of dyes. What is the basic structure of aromatic hydrocarbons?*

The basic structure of an aromatic hydrocarbon is a ring of 6 carbon atoms joined by alternating single and double covalent bonds. This means that within the 6-carbon ring, there are 3 carbon-to-carbon double bonds. The simplest aromatic hydrocarbon is called benzene, C_6H_6. Figure 3–29 shows the structural formula for benzene. Chemists often abbreviate this formula by drawing a hexagon with a circle in the center.

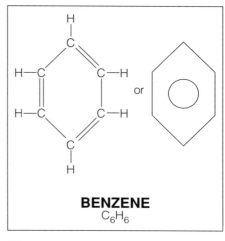

BENZENE
C_6H_6

Figure 3–29 *The structural formula for benzene shows 6 carbon atoms joined by alternating single and double covalent bonds.*

3–5 Section Review

1. What are hydrocarbons? How are hydrocarbons classified?
2. Name three series of hydrocarbons.
3. What is meant by saturated and unsaturated hydrocarbons? Classify each hydrocarbon series according to these definitions.
4. What are aromatic hydrocarbons?

Critical Thinking—*Applying Concepts*
5. Can C_6H_6 be a straight-chain hydrocarbon?

3–6 Substituted Hydrocarbons

Hydrocarbons are but one of several groups of organic compounds. Hydrocarbons contain carbon and hydrogen atoms only. But as you have learned, carbon atoms form bonds with many other elements. So many different groups of organic compounds exist. **The important groups of organic compounds include alcohols, organic acids, esters, and halogen derivatives.** These compounds are called **substituted hydrocarbons.** A substituted hydrocarbon is formed when one or more hydrogen atoms in a hydrocarbon chain or ring is replaced by a different atom or group of atoms.

Alcohols

Alcohols are substituted hydrocarbons in which one or more hydrogen atoms have been replaced by an –OH group, or hydroxyl group. The simplest alcohol is methanol, CH_3OH. You can see from Figure 3–30 that methanol is formed when 1 hydrogen atom in methane is replaced by the –OH group. Methanol is used to make plastics and synthetic fibers. It is also used in automobile gas tank de-icers to prevent water that has condensed in the tank from freezing. Another important use of methanol is as a solvent. Methanol, however, is very poisonous—even when used externally.

As you can tell from the name methanol, alcohols are named by adding the suffix *-ol* to the name of the corresponding hydrocarbon. When an –OH group is substituted for 1 hydrogen atom in ethane, the resulting alcohol is ethanol, C_2H_5OH. Ethanol is produced naturally by the action of yeast or bacteria on the sugar stored in grains such as corn, wheat, and barley.

Ethanol is a good solvent for many organic compounds that do not dissolve in water. Ethanol is used in medicines. It is also the alcohol used in alcoholic beverages. In order to make ethanol available for industrial and medicinal uses only, it must be made unfit for beverage purposes. So poisonous compounds such as methanol are added to ethanol. The resulting mixture is called denatured alcohol.

Figure 3–30 *Methanol, an organic alcohol, is used to make plastics, such as the brightly colored insulation on telephone wires.*

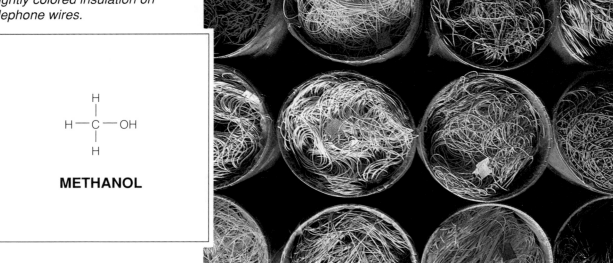

H—C—OH

METHANOL

An alcohol can be in the form of a ring as well as a chain. When 1 hydrogen atom in a benzene ring is replaced by an –OH group, the resulting alcohol is called phenol. Phenol is used in the preparation of plastics and as a disinfectant.

Organic Acids

Organic acids are substituted hydrocarbons that contain the –COOH group, or carboxyl group. Figure 3–32 shows the structural formula for two common organic acids. Notice that one of the carbon-oxygen bonds in the carboxyl group is a double bond.

Organic acids are named by adding the suffix *-oic* to the name of the corresponding hydrocarbon. Most organic acids, however, have common names that are used more frequently. The simplest organic acid is methanoic acid, HCOOH. Methanoic acid is commonly called formic acid. Formic acid is found in nature in the stinging nettle plant and in certain ants. Formic acid produced by an ant causes the ant bite to hurt.

The acid derived from ethane is commonly called acetic acid. Acetic acid is the acid in vinegar. Citric acid, which is found in citrus fruits, is a more complicated organic acid originally derived from the hydrocarbon propane.

Pure
Denatured Alcohol

Shellac thinner
Alcohol stove fuel

DANGER! POISON!
FLAMMABLE. MAY BE FATAL OR CAUSE
BLINDNESS IF SWALLOWED.
VAPOR HARMFUL. EYE IRRITANT.
See other cautions on back panel.

Figure 3–31 *Denatured alcohol is ethanol to which poisonous compounds such as methanol have been added. Why is ethanol denatured?*

Figure 3–32 *Formic acid, also known as methanoic acid, is the simplest organic acid. It is the acid produced by ants and is responsible for the pain caused by an ant bite. Acetic acid is the acid in vinegar. What is another name for acetic acid?*

$$H-\overset{\overset{\textstyle O}{\|}}{C}-OH$$

FORMIC ACID

$$H-\overset{\overset{\textstyle H}{|}}{\underset{\underset{\textstyle H}{|}}{C}}-\overset{\overset{\textstyle O}{\|}}{C}-OH$$

ACETIC ACID

Figure 3–33 *Halogen derivatives have a variety of uses. Polyvinyl chloride is used to make all sorts of water-repellent gear.*

Esters

If an alcohol and an organic acid are chemically combined, the resulting compound is called an ester. Esters are noted for their pleasant aromas and flavors. The substances you read about in the chapter opener that give flavor to ice cream are esters.

Many esters occur naturally. Fruits such as strawberries, bananas, and pineapples get their sweet smell from esters. Esters can also be produced in the laboratory. Synthetic esters are used as perfume additives and as artificial flavorings.

Halogen Derivatives

Hydrocarbons can undergo substitution reactions in which one or more hydrogen atoms are replaced by an atom or atoms of fluorine, chlorine, bromine, or iodine. The family name for these elements is halogens. So substituted hydrocarbons that contain halogens are called halogen derivatives.

A variety of useful substances result from adding halogens to hydrocarbons. The compound methyl chloride, CH_3Cl, is used as a refrigerant. Tetrachloroethane, $C_2H_2Cl_4$, which consists of 4 chlorine atoms substituted in an ethane molecule, is used in dry cleaning.

When 2 hydrogen atoms in a methane molecule are replaced by chlorine atoms and the other 2 hydrogen atoms are replaced by fluorine atoms, a compound commonly known as Freon, CCl_2F_2, is formed. The actual name of this halogen derivative is dichlorodifluoromethane. Freon is the coolant used in many refrigerators and air conditioners.

3–6 Section Review

1. What is a substituted hydrocarbon?
2. What is an alcohol? An organic acid? An ester?

Critical Thinking—*Identifying Relationships*
3. Methanol is used in car de-icers. What does this tell you about the freezing point of methanol? What role does methanol play in terms of the solution of water and methanol?

CONNECTIONS

A Chemical You

The human body is one of the most amazing chemical factories ever created. It can produce chemicals from raw materials, start complex chemical reactions, repair and reproduce some of its own parts, and even correct its own mistakes.

What is the fuel that keeps your human chemical factory going? *Nutrients* contained in the foods you eat maintain the proper functioning of all the systems of the body. The three main types of nutrients—carbohydrates, fats and oils, and proteins—are organic compounds.

Carbohydrates are organic molecules of carbon, hydrogen, and oxygen. Carbohydrates are the body's main source of energy. Carbohydrates can be either sugars or starches. The simplest carbohydrate is the sugar glucose, $C_6H_{12}O_6$. The more complex sugar sucrose, $C_{12}H_{22}O_{11}$, is common table sugar.

Starches are another kind of carbohydrate. Starches are made of long chains of sugar molecules hooked together. Starches are found in foods such as bread, cereal, potatoes, pasta, and rice.

Like carbohydrates, fats and oils contain carbon, hydrogen, and oxygen. These molecules are large, complex esters. Fats and oil store twice as much energy as do carbohydrates.

As a class of organic compounds, fats and oils are sometimes called lipids. Fats are solid at room temperature, whereas oils are liquid. Lipids include cooking oils, butter, and the fat in meat. Although fats and oils are high-energy nutrients, too much of these substances can be a health hazard. Unused fats are stored by the body. This increases body weight. In addition, scientific evidence indicates that eating too much animal fat may contribute to heart disease.

Proteins are used to build and repair body parts. Every living part of your body contains proteins. Blood, muscles, brain tissue, skin, and hair all contain proteins. Proteins contain carbon, hydrogen, oxygen, and nitrogen. Some also contain sulfur and phosphorus. Meat, fish, dairy products, and soybeans are sources of proteins.

A computer-generated model of a human protein molecule

All foods contain three basic nutrients: carbohydrates, fats and oils, and proteins. A balanced diet should provide the proper amounts of each nutrient.

Laboratory Investigation

Acids, Bases, and Salts

Problem

What are some properties of acids and bases?
What happens when acids react with bases?

Materials (per group)

safety goggles	test-tube rack
beaker	red and blue
solutions of	litmus paper
H_2SO_4, HCl,	stirring rod
HNO_3	medicine drop-
solutions of	per
KOH, NaOH,	phenolphthalein
$Ca(OH)_2$	evaporating
6 medium-sized	dish
test tubes	paper towels

Procedure 🧪 🔋 👁

A. *Acids*

1. Put on your safety goggles. Over a sink, pour about 5 mL of each acid into separate test tubes. **CAUTION:** *Handle acids with extreme care. They can burn the skin.* Place the test tubes in the rack. Test the effect of each acid on litmus paper by dipping a stirring rod into the acid and then touching the rod to the litmus paper. Test each acid with both red and blue litmus paper. *Be sure to clean the rod between uses.* Record your observations.

2. Add 1 drop of phenolphthalein to each test tube. Record your observations.

B. *Bases*

1. Over a sink, pour about 5 mL of each base into separate test tubes. **CAUTION:** *Handle bases with extreme care.* Place the test tubes in the rack. Test the contents of each tube with red and blue litmus paper. Record your observations.

2. Add 1 drop of phenolphthalein to each test tube. Record your observations.

3. Place 5 mL of sodium hydroxide solution in a small beaker and add 2 drops of phenolphthalein. Record the solution's color.

4. While slowly stirring, carefully add a few drops of hydrochloric acid until the mixture changes color. Record the color change. This point is known as the indicator endpoint. Test with red and blue litmus paper. Record your observations.

5. Carefully pour some of the mixture into a porcelain evaporating dish. Let the mixture evaporate until it is dry.

Observations

1. What color do acids turn litmus paper? Phenolphthalein?

2. What color do bases turn litmus paper? Phenolphthalein?

3. What happens to the color of the sodium hydroxide-phenolphthalein solution when hydrochloric acid is added?

4. Does the substance formed by the reaction of sodium hydroxide with hydrochloric acid affect litmus paper?

5. Describe the appearance of the substance that remains after evaporation.

Analysis and Conclusions

1. What are some properties of acids? Of bases?

2. What type of substance is formed when an acid reacts with a base? What is the name of this reaction? What is the other product? Why does this substance have no effect on litmus paper?

3. What is meant by an indicator's endpoint?

4. **On Your Own** Write a balanced equation for the reaction between sodium hydroxide and hydrochloric acid.

Summarizing Key Concepts

3–1 Solution Chemistry

▲ A solution consists of a solute and a solvent.

▲ Solubility is a measure of how much of a particular solute can be dissolved in a given amount of solvent at a certain temperature.

▲ Solutions can be unsaturated, saturated, or supersaturated.

3–2 Acids and Bases

▲ Acids taste sour, turn litmus paper red and phenolphthalein colorless, and ionize in water to form hydrogen ions (H^+).

▲ Bases feel slippery, taste bitter, turn litmus paper blue and phenolphthalein pink, and produce hydroxide ions (OH^-) in solution.

3–3 Acids and Bases in Solution: Salts

▲ A neutral substance has a pH of 7. Acids have pH numbers lower than 7. Bases have pH numbers higher than 7.

▲ When an acid chemically combines with a base, the reaction is called neutralization. The products of neutralization are a salt and water.

3–4 Carbon and Its Compounds

▲ Most compounds that contain carbon are called organic compounds.

▲ Isomers have the same molecular formula but different structural formulas.

3–5 Hydrocarbons

▲ The alkanes are saturated hydrocarbons. The alkenes and the alkynes are unsaturated hydrocarbons.

▲ Aromatic hydrocarbons have a ring structure containing 6 carbon atoms.

3–6 Substituted Hydrocarbons

▲ Substituted hydrocarbons include alcohols, organic acids, esters, and halogen derivatives.

Reviewing Key Terms

3–1 Solution Chemistry
solution
solute
solvent
electrolyte
nonelectrolyte
solubility
concentration
concentrated solution
dilute solution
saturated solution
unsaturated solution
supersaturated solution

3–2 Acids and Bases
acid
base

3–3 Acids and Bases in Solution: Salts
pH   neutralization
salt

3–4 Carbon and Its Compounds
organic compound   isomer
structural formula

3–5 Hydrocarbons
hydrocarbon   alkane
saturated hydrocarbon   alkene
unsaturated hydrocarbon   alkyne

3–6 Substituted Hydrocarbons
substituted hydrocarbon

Chapter Review

Content Review

Multiple Choice

Choose the letter of the answer that best completes each statement.

1. A solution that conducts an electric current is called a(an)
   a. nonelectrolyte. c. electrolyte.
   b. tincture.   d. colloid.

2. Which process will not increase the rate of solution of a solid in a liquid?
   a. powdering the solution
   b. cooling the solution
   c. heating the solution
   d. stirring the solution

3. A solution that contains all the solute it can hold at a given temperature is said to be
   a. saturated.   c. supersaturated.
   b. unsaturated.   d. dissociated.

4. Which ion do bases contain?
   a. OH^- b. H_3O^+ c. H^+ d. NH_4^+

5. Organic compounds always contain
   a. carbon.   c. halogens.
   b. oxygen.   d. carboxyl groups.

6. The pH of the products formed by a neutralization reaction is
   a. 1.   b. 7.   c. 14.   d. 0.

7. The type of bonding found in organic compounds is
   a. metallic.   c. covalent.
   b. ionic.   d. coordinate.

8. A compound that contains only carbon and hydrogen is called a(an)
   a. isomer.   c. carbohydrate.
   b. hydrocarbon.   d. alcohol.

9. The simplest aromatic hydrocarbon is
   a. cyclohexane.   c. benzene.
   b. methane.   d. phenol.

10. The –OH group is characteristic of an
    a. organic acid.
    b. aromatic compound.
    c. ester.
    d. alcohol.

True or False

If the statement is true, write "true." If it is false, change the underlined word or words to make the statement true.

1. In a solution, the substance being dissolved is the <u>solvent</u>.
2. Acids are often defined as <u>proton donors</u>.
3. Strong acids are <u>poor</u> electrolytes.
4. A neutral solution has a pH of <u>10</u>.
5. In <u>neutralization</u>, an acid reacts with a base.
6. Hydrocarbons that contain only single bonds are said to be <u>unsaturated</u>.
7. The 4-carbon alkane is called <u>butane</u>.
8. An organic acid is characterized by the group <u>–COOH</u>.
9. Compounds that have the same molecular formula but different structural formulas are called <u>isotopes</u>.

Concept Mapping

Complete the following concept map for Section 3–1. Refer to pages O6–O7 to construct a concept map for the entire chapter.

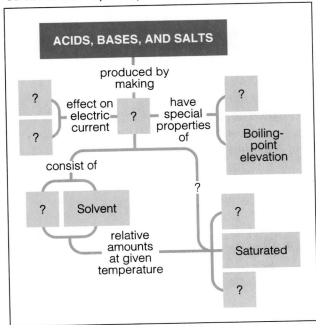

Concept Mastery

Discuss each of the following in a brief paragraph.

1. Describe three ways in which the rate of solution of a solid in a liquid can be increased. Use a specific example.
2. Explain how a saturated solution of one solute can be concentrated and a saturated solution of a different solute can be dilute.
3. Explain why water is both an acid and a base.
4. Discuss four reasons why carbon compounds are so abundant.
5. Explain the importance of structural formulas in organic chemistry.
6. In this chapter you learned that compounds called bases contain the hydroxide ion, $-OH^-$. Alcohols also contain the $-OH$ group. Why are alcohols not bases?

Critical Thinking and Problem Solving

Use the skills you have developed in this chapter to answer each of the following

1. **Classifying compounds** Identify each of the following compounds as an acid, base, or salt:
   a. $CaCO_3$
   b. HI
   c. $CsOH$
   d. H_3PO_4
   e. $MgSO_4$
   f. $Ga(OH)_3$
2. **Applying concepts** The caps are removed from a warm bottle and a cold bottle of carbonated beverage. The soda in the cold bottle fizzes slightly. The soda in the warm bottle fizzes rapidly.
   a. What two conditions affecting solubility are present here?
   b. Which condition is the variable?
   c. Give an explanation for what happens.
3. **Designing an experiment** Describe an experiment to determine whether a solution is saturated or unsaturated.
4. **Identifying patterns** A crystal of solute is added to a saturated, unsaturated, and supersaturated solution. Describe what happens in each case.
5. **Classifying hydrocarbons** Classify each of the following hydrocarbons as an alkane, alkene, alkyne, or aromatic compound.
   a. C_4H_{10}
   b. C_3H_4
   c. C_2H_4
   d. C_6H_6
   e. $C_{13}H_{26}$
   f. $C_{42}H_{82}$
6. **Applying definitions** Draw the structural formulas for the isomers of hexane.
7. **Making diagrams** Draw structural formulas for the following compounds:
   a. hexane
   b. butene
   c. propyne
8. **Classifying substituted hydrocarbons** Classify each of the following compounds as a(an) ester, alcohol, organic acid, or halogen derivative:
   a. $C_2H_5COOC_3H_7$
   b. C_4H_9Cl
   c. C_6H_5OH
   d. C_2H_5COOH
9. **Identifying patterns** Using the general formulas for the alkanes, alkenes, and alkynes, show why the number of hydrogen atoms in each series decreases by two.
10. **Drawing a conclusion** Explain why the alkene series and the alkyne series begin with a 2-carbon hydrocarbon rather than a 1-carbon hydrocarbon, as the alkane series does. Use structural formulas to support your explanation.
11. **Using the writing process** Water is often called the universal solvent. Is it truly? Write a short story about whether a true universal solvent exists. Describe all its physical properties. End your story by describing how you would bring a sample of it to class.

Petrochemical Technology

A telephone call at any time of the day or night may summon you to travel to a distant location. Once you arrive at your destination, your life will be in danger almost constantly. The situation will be explosive. One wrong move—even one unfortunate act of nature—could result in a tremendous and fatal explosion. Your role there may last only a day or perhaps several months. What is this awesome-sounding job? It is putting out oil-well fires. Does it sound like a job you might enjoy? Probably not. But one brave and unusual man who goes by the name of Red Adair loves it!

Oil wells, and the liquids and gases within them, are highly flammable. Once an oil fire begins, it will burn until all the fuel is gone. Putting out an oil-well fire requires enormous ingenuity and courage. Nonetheless, Adair has never failed to put out a fire. Some have taken six months and others only thirty seconds.

What is so important about oil that people are willing to risk their lives to find it, extract it from the Earth, and protect it? In this chapter you will learn the answer to that question. You will also find out what oil is and how it can be used.

Journal *Activity*

You and Your World Look around your home at the objects that surround you every day. Do you see a lot of plastic? You can probably find plastic in your closet, on your shelves, in your refrigerator—anywhere you look. In our journal, describe some of the objects you find that are made of plastic. Explain why the plastic is useful and what material may have been used before plastic was developed.

◀ *Only the know-how of Red Adair (inset) is enough to match an oil field burning out of control.*

Guide for Reading

Focus on these questions as you read.

▶ What is petroleum?

▶ How does fractional distillation separate petroleum into its various components?

ACTIVITY

DOING

Where Is the Oil?

Using books and other reference materials in the library, find out where the major oil fields in the world are located. Make a map showing these oil fields. Be sure to include oil fields that have recently been discovered, as well as areas that are currently believed to contain oil.

4–1 What Is Petroleum?

Have you ever seen a movie in which a group of people gathered around a well were shouting and cheering as a thick black liquid was rising up through the ground and spurting high into the air? Or have you ever read about geologists who study the composition of the Earth in an attempt to find oil? If so, you may have realized—correctly so—that crude oil is a rather valuable substance. This sought-after crude oil, one form of **petroleum,** has been called black gold because of its tremendous importance. Fuels made from petroleum provide nearly half the energy used in the world. And thousands of products—from the bathing suit you wear when swimming to the toothpaste you use when brushing your teeth—are made from this petroleum.

Petroleum is a substance believed to have been formed hundreds of millions of years ago when layers of dead plants and animals were buried beneath sediments such as mud, sand, silt, or clay at the bottom of the oceans. Over millions of years, heat and great pressure changed the plant and animal remains into petroleum. Petroleum is a nonrenewable resource. A nonrenewable resource is one that

Figure 4–1 *The sight of thick black liquid shooting high into the air has been a welcomed one since the first oil well was built in Pennsylvania.*

cannot be replaced once it is used up. There is only a certain amount of petroleum in existence. Once the existing petroleum is used up, no more will be available.

Despite the huge variety of products obtained from petroleum, few people ever see the substance itself. The liquid form that gushes from deep within the Earth is a mixture of chemicals called crude oil. Petroleum can also be found as a solid in certain rocks and sand. It has been called black gold because it is usually black or dark brown. But it can be green, red, yellow, or even colorless. Petroleum may flow as easily as water, or it may ooze slowly—like thick tar. The color and thickness of petroleum depend on the substances that make it up.

Separating Petroleum Into Parts

By itself, petroleum is almost useless. But the different parts, or **fractions,** of petroleum are among the most useful chemicals in the world. **Petroleum is separated into its useful parts by a process called fractional distillation.** The process of distillation involves heating a liquid until it vaporizes (changes into a gas) and then allowing the vapor to cool until it condenses (turns back into a liquid). The different fractions of petroleum have different boiling points. So each fraction vaporizes at a different temperature than do the others. The temperature at which a substance boils is the same as the temperature at which it condenses. So if each fraction vaporizes at a different temperature, then each fraction will condense back to a liquid at a different temperature. By removing, or drawing off, each fraction as it condenses, petroleum can easily be separated into its various parts.

Fractional distillation of petroleum is done in a fractionating tower. The process of separating petroleum into its fractions is called **refining.** Refining petroleum is done at a large plant called a refinery. At a refinery, fractionating towers may rise 30 meters or more. Figure 4–3 on page 100 shows a fractionating tower. Petroleum is piped into the base of the fractionating tower and heated to about 385°C. At this temperature, which is higher than the boiling points of most of the fractions, the petroleum vaporizes.

PRODUCTS PRODUCED
FROM A BARREL OF CRUDE OIL

	% Yield
Gasolines	46.7
Fuel oil	28.6
Jet fuel	9.1
Petrochemicals and Miscellaneous products	3.8
Coke	3.5
Asphalt and road oil	3.1
Liquified gases	2.9
Lubricants	1.3
Kerosene	0.9
Waxes	0.1

Figure 4–2 *Petroleum is separated into fractions in fractionating columns in an oil refinery. The numbers inside the barrel show the amount, or percentage yield, of each fraction that can be obtained from a barrel of crude oil. Which fraction represents the highest percentage yield?*

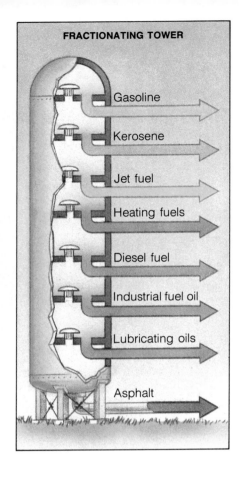

FRACTIONATING TOWER

- Gasoline
- Kerosene
- Jet fuel
- Heating fuels
- Diesel fuel
- Industrial fuel oil
- Lubricating oils
- Asphalt

Figure 4–3 Each fraction of petroleum condenses at a different temperature and is drawn off in collecting vessels located at fixed points along the column. Where in the tower is the temperature lowest? Where is it highest?

When the petroleum vaporizes, the fractions rise up the tower. As they rise, they cool and condense. Some fractions condense at high temperatures. These fractions condense right away near the bottom of the tower and are drawn off into collecting vessels. Other fractions continue to rise in the tower. These fractions are drawn off at higher levels in the tower. As a result of this vaporization-condensation process, the various fractions of petroleum are separated and collected.

You will notice in Figure 4–3 that asphalt is collected at the bottom of the fractionating tower. Asphalt requires a temperature even higher than 385°C to vaporize. When the other fractions vaporize, asphalt is left behind as a liquid that runs out of the bottom of the tower. Which fraction in the tower condenses at the lowest temperature?

Petroleum Products

Asphalt—the main material used for building roads—is one product that comes directly from petroleum. Wax, used in furniture polish and milk cartons, is another. Asphalt and wax fall into the category of raw materials that come from the separation of petroleum and are used in manufacturing. Many of the other raw materials in this category, however, are converted to chemicals from which a variety of products—ranging from cosmetics to fertilizers—are made. You will read about these products in the next section.

Another group of petroleum products includes lubricants. Lubricants are substances that reduce friction between moving parts of equipment. The oil applied to the gears of a bicycle is an example of a lubricant. Lubricants are used in many machines—from delicate scientific equipment to the landing gear of an aircraft. Can you think of some other uses of lubricants?

Figure 4–4 The asphalt used to pave roadways comes directly from petroleum. Lubricants used to make machinery run more efficiently also come from petroleum.

The greatest percentage of petroleum products includes fuels. Fuels made from petroleum burn easily and release a tremendous amount of energy, primarily in the form of heat. They are also easier to handle, store, and transport than are other fuels, such as coal and wood. Petroleum is the source of nearly all the fuels used for transportation and the many fuels used to produce heat and electricity.

4–1 Section Review

1. What is petroleum and why is it important?
2. Describe the process of fractional distillation.
3. Describe the products produced from petroleum.

Critical Thinking—*Applying Concepts*
4. How would you separate three substances—A, B, and C—whose boiling points are 50°C, 100°C, and 150°C, respectively?

ACTIVITY READING

All About Oil

Some discoveries in history are acts of genius. Others are accidents. But most are a little of both. Read Isaac Asimov's *How Did We Find Out About Oil?* to discover how oil was first discovered, extracted, and used.

Activity Bank

Oil Spill, p.157

4–2 Petrochemical Products

Paint a picture, pour milk from a plastic container, or put on a pair of sneakers and you are using a product made from petroleum, or a **petrochemical product.**

Polymer Chemistry

The petrochemical products that are part of your life come from the chemicals produced from petroleum. (The word petrochemical refers to chemicals that come from petroleum.) Petrochemical products usually consist of molecules that take the form of long chains. Each link in the chain is a small molecular unit called a **monomer.** (The prefix *mono-* means singular, or one.) The entire molecule chain is called a **polymer.** (The prefix *poly-* means many.)

Guide for Reading

Focus on these questions as you read.

▶ What is a petrochemical product?

▶ What are some products of polymerization?

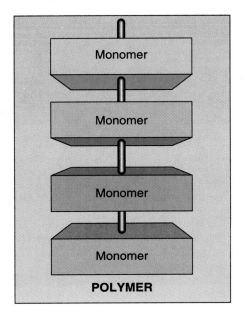

Figure 4–5 *A polymer is made up of a series of monomers. What factors distinguish one polymer from another?*

Figure 4–6 *Silk is spun from threads of silk worm cocoons. Silk is a natural polymer. Rubber obtained from rubber trees is also a natural polymer.*

The types of monomers and the length and shape of the polymer chain determine the physical properties of the polymer. Manufacturers of petrochemical products join monomers together to build polymers. A general term for this process is polymer chemistry.

Natural Polymers

Most of the polymers you will read about in this chapter are made from petrochemicals. Some polymers, however, do occur in nature. Cotton, silk, wool, and natural rubber are all **natural polymers.** Cellulose and lignin, which are important parts of wood, are natural polymers. In fact, all living things contain polymers. Yes—that includes you! Protein, an essential ingredient of living matter, is a polymer. The monomers from which proteins are made are called amino acids. Combined in groups of one hundred or more units, amino-acid monomers form many of the parts of your body—from hair to heart muscle.

Synthetic Polymers

The first polymer was manufactured in 1909. Since then, **polymerization** (poh-lihm-er-uh-ZAY-shuhn) has come a long way. Polymerization is the process of chemically bonding monomers to form polymers. Most early polymers consisted of fewer than two hundred monomers. Today's polymers may contain thousands of monomers. The many ways in which these monomers can be linked may be very complex. They include single chains, parallel chains, intertwining chains, spirals, loops, and loops of chains!

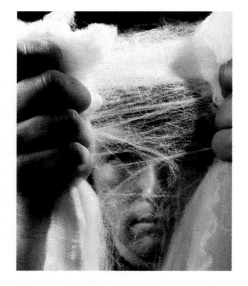

Figure 4–7 *Synthetic fibers, such as nylon, are used to make the carpets you walk on. What are some other uses of synthetic fibers?*

Figure 4–8 *Applications of synthetic polymers include rubber tires, aspirin, and waterproof rain gear.*

Polymers produced from petrochemicals are called **synthetic polymers.** Something that is synthetic does not exist naturally. Instead it is made by people. Polymer chemistry has produced synthetic materials that are strong, lightweight, heat resistant, flexible, and durable (long lasting). These properties give polymers a wide range of applications.

Although the term polymer may be new to you, you will soon discover that many polymers produced from petrochemicals are familiar to you. For example, petrochemical products such as synthetic rubber and plastic wrap are synthetic polymers. Synthetic polymers are used to make fabrics such as nylon, rayon, Orlon, and Dacron. Plastics—used in products from kitchen utensils to rocket engines—are petrochemical products made of polymers.

In medicine, polymers are used as substitutes for human tissues, such as bones and arteries. These polymers must last a lifetime and must withstand the wear and tear of constant use. Polymer adhesives, rather than thread, may be used to hold clothes together. Polymers are replacing glass, metal, and paper as containers for food. The cup of hot chocolate you may have held today did not burn your hand because it was made of a white insulating polymer. Polymer materials are also used to make rugs, furniture, wall coverings, and curtains. Look around and see how many polymers you can spot. And remember: You have petroleum to thank for all these useful materials.

Polymer materials also can be mixed and matched to produce substances with unusual properties.

ACTIVITY

DOING

Homemade Adhesives

1. Mix flour and water to make a paste.

2. Separate the white from the yolk of a raw egg. Egg white is a natural adhesive.

Compare the "sticking strength" of the flour paste and the egg white by using each to lift objects of increasing mass.

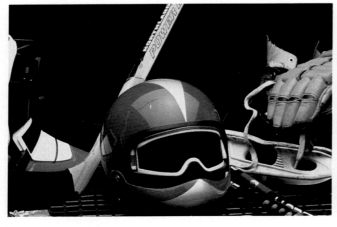

Figure 4–9 *Different forms of plastics are made by melting and processing small bits of plastic grain. The plastic grain can be made into a variety of products, each with characteristics suited for a specific use.*

Different plastics and synthetic fibers are combined to make puncture-proof tires and bulletproof vests. Layers of polymer materials can be combined to make waterproof rain gear.

Polymer chemistry is also important in the transportation industry. Every year the number of polymer parts in cars, planes, and trains increases. A plastic car engine has been built and tested. This engine is lighter, more fuel efficient, and more durable than a metal engine. As you can see, polymers made from petroleum are extremely important today. And they will be even more important in the future.

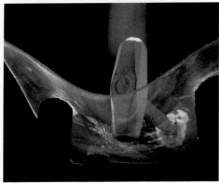

Figure 4–10 *Polymer technology has produced some amazing materials. This sheet of plastic bends no matter how hard it is struck by a hammer (bottom). This extra-thin sheet of plastic is not damaged by temperatures greater than 1000°C (top).*

4–2 Section Review

1. What is a petrochemical product? Give several examples.
2. What is the relationship between a monomer and a polymer?
3. List three examples of natural polymers.
4. What is polymerization?
5. What are some characteristics of synthetic polymers?

Connection—*You and Your World*

6. What might be some of the economic side effects of increased use of polymers in automobiles?

CONNECTIONS

Life-Saving Chemistry

The human body is an amazing collection of systems that interact with one another to sustain life. The systems are made up of tissues and cells that each have a certain function. For example, thin, soft, flexible layers of tissue control the flow of chemicals into and out of the cells. These layers are called *biological membranes.* Some biological membranes allow harmful chemicals to pass out of a cell while keeping needed chemicals inside the cell. The kidneys contain membranes that do this kind of job. Biological membranes are natural polymers.

Like other parts of the human body, biological membranes sometimes do not work correctly or become damaged. The failure of a biological membrane to perform its necessary function can be fatal. That is why healthy kidneys are so important. In the human body, two kidneys constantly filter waste materials from the blood. If the kidneys fail to do this vital job, wastes will remain in the blood, causing damage to other parts of the body.

In 1944, a synthetic membrane that worked like a natural biological membrane was developed. This synthetic membrane made possible the invention of the artificial kidney. The artificial kidney is essentially a large filtering machine. A tube from the machine is inserted into an artery in the patient's arm. Blood flows from the patient into the machine, where the blood is filtered by a synthetic membrane. The purified blood is then pumped back into the patient's system.

Synthetic membranes have a variety of other applications. Some are used as skin substitutes for people who have been badly burned. Others are used in artificial body parts. Polymer chemistry has had a significant impact on human health.

No, the bird is not drowning. It is surrounded by a synthetic membrane that allows oxygen to pass from the water to the bird.

Laboratory Investigation

Comparing Natural and Synthetic Polymers

Problem

How do natural and synthetic polymers compare in strength, absorbency, and resistance to chemical damage?

Materials *(per group)*

3 samples of natural polymer cloth:
   wool, cotton, linen
3 samples of synthetic polymer cloth:
   polyester, nylon, acetate
12 Styrofoam cups
mild acid (lime or lemon juice or vinegar)

marking pen	liquid bleach
metric ruler	medicine dropper
scissors	oil
rubber gloves	paper towel

Procedure 🧪 📷 👁 ⚖

1. Record the color of each cloth.
2. Label 6 Styrofoam cups with the names of the 6 cloth samples. Also write the word Bleach on each cup.
3. Cut a 2-square-cm piece from each cloth. Put each piece in its cup.
4. Wearing rubber gloves, carefully pour a small amount of bleach into each cup.
5. Label the 6 remaining cups with the names of the 6 cloth samples and the word Acid. Then pour a small amount of the mild acid into each and repeat step 3.

6. Set the cups aside for 24 hours. Meanwhile, proceed with steps 7 through 9.
7. Using the remaining samples of cloth, attempt to tear each.
8. Place a drop of water on each material. Note whether the water forms beads or is absorbed. If the water is absorbed, record the rate of absorption.
9. Repeat step 8 using a drop of oil.
10. After 24 hours, wearing rubber gloves, carefully pour the liquids in the cups into the sink or a container provided by your teacher. Dry the samples with a paper towel.
11. Record any color changes.

Observations

1. Which material held its color best in bleach? In acid?
2. Which materials were least resistant to chemical damage by bleach or mild acid?
3. Which material has the strongest fiber or is hardest to tear?
4. Which materials are water repellent?

Analysis and Conclusions

1. Compare the natural and synthetic polymers' strength, absorbency, and resistance to chemical change.
2. Which material would you use to manufacture a laboratory coat? A farmer's overalls? A raincoat? An auto mechanic's shirt?
3. **On Your Own** Confirm your results with additional samples of natural and synthetic polymers. What additional tests can you add for comparison?

Study Guide

Summarizing Key Concepts

4–1 What Is Petroleum?

▲ Petroleum is a substance believed to have been formed when plant and animal remains were subjected to tremendous pressure for millions of years.

▲ Petroleum is a mixture of chemicals that can be divided into separate parts, or fractions.

▲ Petroleum is separated into its components through a process call fractional distillation.

▲ During fractional distillation, petroleum is heated and pumped into a fractionating tower. As the fractions rise up the tower, they cool and condense. Because they condense at different temperatures, they can be drawn off at different heights.

▲ Different groups of products are produced from petroleum: raw materials used in manufacturing, raw materials converted to chemicals, lubricants, and fuels.

4–2 Petrochemical Products

▲ Substances derived from petroleum are called petrochemical products.

▲ Most petrochemical products are polymers.

▲ A polymer is a series of molecular units called monomers.

▲ The process of chemically combining monomers to make a polymer is called polymerization.

▲ Natural polymers include cotton, silk, wool, natural rubber, cellulose, protein, and lignin.

▲ Synthetic polymers include synthetic rubber, plastics, and fabrics such as nylon, Orlon, rayon, and Dacron.

▲ Polymers are usually strong, lightweight, heat resistant, flexible, and durable.

▲ Polymers can be mixed and matched to form substances that are waterproof, puncture proof, or electrically conductive.

Reviewing Key Terms

Define each term in a complete sentence.

4–1 What Is Petroleum?
petroleum
fraction
refining

4–2 Petrochemical Products
petrochemical product
monomer
polymer
natural polymer
polymerization
synthetic polymer

Chapter Review

Content Review

Multiple Choice

Choose the letter of the answer that best completes each statement.

1. Crude oil is
   a. a single element.   c. asphalt.
   b. gasoline.   d. a mixture.
2. The physical property used to separate petroleum into its parts is
   a. melting point.   c. density.
   b. boiling point.   d. solubility.
3. The highest temperature in the fractionating tower is
   a. below the boiling point of most petroleum fractions.
   b. above the boiling point of most petroleum fractions.
   c. equal to the boiling point of most petroleum fractions.
   d. below the melting point of most petroleum fractions.
4. A substance unlikely to vaporize in a fractionating tower is
   a. kerosene.   c. asphalt.
   b. gasoline.   d. heating fuel.

5. The process of distillation involves
   a. vaporization and condensation.
   b. freezing and melting.
   c. vaporization and melting.
   d. freezing and condensation.
6. A polymer is made of a series of
   a. atoms.   c. fuels.
   b. monomers.   d. synthetic molecules.
7. An example of a natural polymer is
   a. wool.   c. crude oil.
   b. plastic.   d. copper.
8. An example of a synthetic polymer is
   a. natural rubber.   c. rayon.
   b. protein.   d. cotton.
9. The process of chemically bonding monomers to form polymers is called
   a. distillation.   c. polymerization.
   b. fractionation.   d. refining.
10. An example of a polymer product is
   a. lead tubing.   c. water.
   b. crude oil.   d. plastic.

True or False

If the statement is true, write "true." If it is false, change the underlined word or words to make the statement true.

1. Petroleum taken directly from the Earth is called <u>asphalt</u>.
2. Petroleum can be separated into its different parts, or <u>fractions</u>.
3. The process of separating petroleum into its components is called <u>condensing</u>.
4. <u>Polymerization</u> involves the chemical bonding of monomers into polymers.
5. Plastics are examples of <u>natural</u> polymers.
6. When a vapor <u>evaporates</u>, it changes back to a liquid.
7. A <u>monomer</u> is a long chain of <u>polymers</u>.
8. Silk is an example of a <u>natural</u> polymer.

Concept Mapping

Complete the following concept map for Section 4–1. Refer to pages O6–O7 to construct a concept map for the entire chapter.

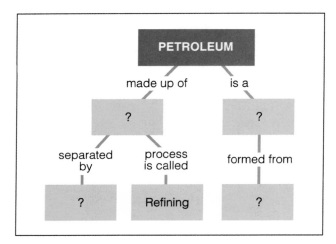

Concept Mastery

Discuss each of the following in a brief paragraph.

1. How did petroleum most likely form?
2. Explain how petroleum is separated into parts during fractional distillation.
3. What happens in the fractionating tower to fractions with extremely high boiling points? With extremely low boiling points?
4. Why is petroleum an important source of fuel?
5. What are natural polymers? Give some examples.
6. What is polymerization? Why is this process useful?
7. What are synthetic polymers? Give some examples.
8. What characteristics of synthetic polymers make them so useful?

Critical Thinking and Problem Solving

Use the skills you have developed in this chapter to answer each of the following.

1. **Drawing conclusions** Petroleum is believed to have been formed at the bottom of oceans. However, oil is often found under dry land—even in deserts. How can you explain this?

2. **Relating facts** Describe some of the uses of synthetic polymers in your life. Then describe what changes you would have to make in your lifestyle if these polymers were not available.
3. **Identifying patterns** The achievements and inventions of the United States space program—originally directed outside planet Earth—have found dramatic uses in our everyday lives. Why do you think space-technology spinoffs have been so widely and successfully used?
4. **Applying concepts** Describe several ways in which polymer chemistry can be used in medicine other than those mentioned in the chapter.
5. **Analyzing information** The United States has been said to have undergone a chemical revolution during the last fifty years. Having read this chapter, explain what the term chemical revolution means to you.
6. **Applying technology** Imagine that you are an engineer whose task is to design a fuel-efficient car that meets all the current standards for safety and durabilty. What kinds of materials would you consider using? What properties must the materials used in the engine have? How about the materials used in safety belts and seat cushions?
7. **Using the writing process** Many people use and enjoy the products derived from petroleum. However, some of the chemical processes used to make these products—as well as the burning of petroleum fuels—add to the pollution of the air, land, and water. Should the manufacturing of certain products be stopped in order to protect the environment? Write a composition expressing your opinion and explaining the reason for it.

Radioactive Elements

Guide for Reading

After you read the following sections, you will be able to

5–1 Radioactivity
- Define radioactivity.
- Compare the three types of nuclear radiation.

5–2 Nuclear Reactions
- Describe the process and products of radioactive decay.
- Describe artificial transmutation.

5–3 Harnessing the Nucleus
- Compare nuclear fission and nuclear fusion.

5–4 Detecting and Using Radioactivity
- Identify instruments that can detect and measure radioactivity.
- Discuss ways in which radioactive substances can be used.

The rich prairie land near Waxahachie, Texas, is the site of one of the most ambitious scientific research projects ever undertaken. The project will take many years and billions of dollars to complete. Yet even when construction ends, it will appear that not much has been done. Why? This project is being built almost 46 meters beneath the soil.

The project is the world's largest accelerator, known as the Superconducting Supercollider (SSC). The supercollider will be housed in a circular tunnel some 85 kilometers in length! Protons will shoot around and around the tunnel, controlled by more than 9000 superconducting magnets that will keep the protons on track and help boost their speed.

When the protons are traveling at almost the speed of light, they will be diverted toward the nuclei of target atoms. The reactions that occur during the collisions of protons and target nuclei hold long sought-after information.

Why is smashing tiny particles into atomic nuclei worth all the effort? In this chapter you will discover the importance of subatomic particles, their relation to the nucleus of an atom, and their role in producing enormous quantities of energy.

Journal *Activity*

You and Your World You have probably heard the terms atomic bomb (A-bomb) and hydrogen bomb (H-bomb). Perhaps you even know about the nuclear radiation associated with them. In your journal, describe the thoughts and/or questions that come to mind when you hear these terms.

Studying the tracks made by subatomic particles enables scientists to learn more about matter and energy.

5–1 Radioactivity

Some discoveries are made by performing experiments to find out whether hypotheses are true. Other discoveries are stumbled upon purely by accident. The majority of scientific discoveries, however, are a combination of the two—both genius and luck. One such discovery was made by the French scientist Henri Becquerel (beh-KREHL) in 1896. Becquerel was experimenting with a uranium compound to determine whether it gave off X-rays. His experiments did indeed provide evidence of X-rays. But they also showed something else—something rather exciting. Quite by accident, Becquerel discovered that the uranium compound gave off other types of rays that had never before been detected. Little did Becquerel know then that these mysterious rays would open up a whole new world of modern science.

An Illuminating Discovery

At the time of Becquerel's work, scientists knew that certain substances glowed when exposed to sunlight. Such substances are said to be fluorescent. Becquerel wondered whether in addition to glowing, fluorescent substances gave off X-rays.

To test his hypothesis, Becquerel wrapped some photographic film in lightproof paper (paper that does not allow light through it). He placed a piece

Figure 5–1 *The image on the photographic film (right) convinced Becquerel that an invisible "something" had been given off by uranium. The photograph on the left shows rectangular blocks containing the element cesium. The photograph was taken through a heavy glass window 1 meter thick with no source of illumination other than the cesium.*

of fluorescent uranium salt on top of the film and the paper and set both in the sun. Becquerel reasoned that if X-rays were produced by the uranium salt when it fluoresced, the X-rays would pass through the lightproof paper and produce an image on the film. The lightproof paper would prevent light from reaching the film and creating an image, however.

When Becquerel developed the film, he was delighted to see an image. The image was evidence that fluorescent substances give off X-rays when exposed to sunlight. In order to confirm his results, he prepared another sample of uranium salt and film to repeat his experiment the following day. But much to his disappointment, the next two days were cloudy. Impatient to get on with his work, Becquerel decided to develop the film anyway. What he saw on the film amazed him. Once again there was an image of the sample, even though the uranium salt had not been made to fluoresce. In fact, the image on the film was just as strong and clear as the image that had been formed when the sample was exposed to sunlight.

Becquerel realized that an invisible "something" given off by the salt had gone through the lightproof paper and produced an image. In time, this invisible "something" was named **nuclear radiation.** Becquerel tested many more uranium compounds and concluded that the source of nuclear radiation was the element uranium. An element that gives off nuclear radiation is said to be **radioactive.**

Marie Curie, a Polish scientist working in France and a former student of Becquerel's, became interested in Becquerel's pioneering work. She and her husband, French scientist Pierre Curie, began searching for other radioactive elements. In 1898, the Curies discovered a new radioactive element in a uranium ore known as pitchblende. They named the element polonium in honor of Marie Curie's native Poland. Later that year they discovered another radioactive element. They named this element radium, which means "shining element." Both polonium and radium are more radioactive than uranium. Since the Curies' discovery of polonium and radium, many other radioactive elements have been identified and even artificially produced.

Figure 5–2 *Marie Curie and her husband, Pierre, were responsible for the discovery of the radioactive elements radium and polonium. Since that time, many other radioactive elements have been identified.*

The Nature of Nuclear Radiation

Nuclear radiation cannot be seen. So radioactive elements were difficult to identify at first. But it was quickly realized that radioactive elements have certain characteristic properties. The first of these is the property observed by Becquerel. Nuclear radiation given off by radioactive elements alters photographic film. Another property of many radioactive elements is that they produce fluorescence in certain compounds. A third characteristic is that electric charge can be detected in the air surrounding radioactive elements. Finally, nuclear radiation damages cells in most organisms.

Today, scientists use the term **radioactivity** to describe the phenomenon discovered by Becquerel. **Radioactivity is the release of nuclear radiation in the form of particles and rays from a radioactive element.** The radiation given off by radioactive elements consists of three different particles or rays. The three types of radiation have been named alpha (AL-fuh) particles, beta (BAYT-uh) particles, and gamma (GAM-uh) rays after the first three letters of the Greek alphabet.

Figure 5–3 *The three types of radiation can be separated according to charge and penetrating power. When passed through a magnetic field, alpha particles are deflected toward the negative magnetic pole, beta particles are deflected toward the positive pole, and gamma rays are not deflected. Which type of radiation is the most penetrating?*

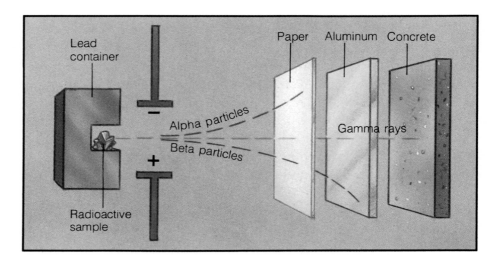

Figure 5–4 *This chart shows the three types of nuclear radiation. What is an alpha particle? A beta particle? Describe gamma rays.*

TYPES OF NUCLEAR RADIATION		
Type	Atomic Mass	Atomic Number
Alpha (α)	4	2
Beta (β)	0	–1
Gamma (γ)	0	none

ALPHA PARTICLES An **alpha particle** is actually the nucleus of a helium atom—2 protons and 2 neutrons. An alpha particle has a positive charge because it contains 2 positive protons and no other charges. Alpha particles are the weakest type of nuclear radiation. Although they can burn flesh, alpha particles can be stopped by a sheet of paper.

BETA PARTICLES A **beta particle** is an electron. However, a beta particle should not be confused with an electron that surrounds the nucleus of an atom. A beta particle is an electron that is formed inside the nucleus when a neutron breaks apart. Beta particles have a penetrating ability 100 times greater than alpha particles. Beta particles can pass through as much as 3 millimeters of aluminum.

GAMMA RAYS A **gamma ray** is an electromagnetic wave of extremely high frequency and short wavelength. Gamma rays are the same kind of waves as the visible light that enables you to see. That is, both are forms of electromagnetic waves. Gamma rays, however, carry a lot more energy. They are the most penetrating radiation given off by radioactive elements. Gamma rays can pass through several centimeters of lead!

5–1 Section Review

1. Describe radioactivity. How did Becquerel discover radioactivity?
2. How did the Curies use Becquerel's discovery?
3. What is a radioactive element?
4. Describe an alpha particle, a beta particle, and a gamma ray. How are they alike? How are they different?

Critical Thinking—*Forming a Hypothesis*
5. There are several theories that attempt to explain how a beta particle is produced. Develop a hypothesis to explain what a neutron—since it is neutral—may actually be composed of while still containing an electron.

Guide for Reading

Focus on these questions as you read.

▶ Why do radioactive nuclei undergo nuclear reactions?

▶ How are radioactive decay and artificial transmutation similar and how are they different?

5–2 Nuclear Reactions

Although Becquerel and the Curies observed radioactivity, they could not explain its origin. The reason for this is understandable: The source of radioactivity is the nucleus of an atom. But Becquerel discovered radioactivity well before the nucleus was discovered. Several years after Becquerel's and the Curies' work, it was determined that radioactivity results when the nuclei of atoms of certain elements change, emitting particles and/or rays. What still remained unknown, however, was what makes a nucleus break apart and why only some elements are radioactive.

Nuclear Stability

The answers to these puzzling questions would be found in the atom—specifically, in the nucleus. The nucleus of an atom contains protons and neutrons. Protons are positively charged particles. Neutrons are neutral particles; they have no charge. It is a scientific fact that particles with the same charge (positive or negative) repel each other. Thus protons repel each other. How, then, does the nucleus hold together? A force known as the **nuclear strong force** overcomes the force of repulsion between protons and holds protons and neutrons together in the nucleus. The energy associated with the strong force is called **binding energy.**

The binding energy is essential to the stability of a nucleus. In some atoms, the binding energy is great enough to hold the nucleus together permanently. The nuclei of such atoms are said to be stable. In other atoms, the binding energy is not as great. The nuclei of these atoms are said to be unstable. An unstable nucleus will come apart. Atoms with unstable nuclei are radioactive.

Some elements that are not radioactive have radioactive forms, or **isotopes** (IGH-suh-tohps). What is

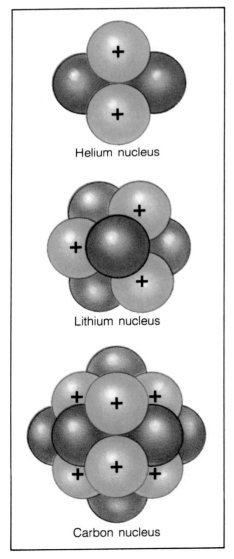

Helium nucleus

Lithium nucleus

Carbon nucleus

Figure 5–5 *The nucleus of an atom contains positively charged protons and neutral neutrons that are held together by the nuclear strong force. A helium nucleus has 2 protons and 2 neutrons. How many protons and neutrons does a lithium nucleus contain? A carbon nucleus?*

NONRADIOACTIVE AND RADIOACTIVE ISOTOPES OF SOME COMMON ELEMENTS

Element	Nonradioactive Isotope	Radioactive Isotope
Hydrogen	1 proton 0 neutrons	1 proton 2 neutrons
Helium	2 protons 2 neutrons	2 protons 4 neutrons
Lithium	3 protons 4 neutrons	3 protons 5 neutrons
Carbon	6 protons 6 neutrons	6 protons 8 neutrons
Nitrogen	7 protons 7 neutrons	7 protons 9 neutrons
Oxygen	8 protons 8 neutrons	8 protons 6 neutrons
Potassium	19 protons 20 neutrons	19 protons 21 neutrons

Figure 5–6 *An isotope is an atom of an element that has the same number of protons but a different number of neutrons. Often, if the number of neutrons greatly differs from the number of protons, the isotope does not have enough binding energy to hold the nucleus together and is therefore radioactive.*

an isotope? The number of protons in the atoms of a particular element cannot vary. An atom is identified by the number of protons it contains. (The number of protons is called the atomic number.) Carbon atoms would not be carbon atoms if they had 5 protons or 7 protons—only 6 protons will do. Yet there are some carbon atoms that have 6 neutrons, and others that have 8 neutrons. The difference in the number of neutrons affects the characteristics of the atom but not its identity. Atoms that have the same number of protons (atomic number) but different numbers of neutrons are called isotopes.

Many elements have at least one radioactive isotope. For example, carbon has two common isotopes—carbon-12 and carbon-14. Carbon-12, which you are familiar with as coal, graphite, and diamond, is not radioactive. Carbon-14, used in dating fossils, is radioactive. Figure 5–6 shows the radioactive and nonradioactive isotopes of some common elements.

ACTIVITY

WRITING

What Is a Quark?

Scientists have proposed that all nuclear subatomic particles are composed of basic particles called quarks. Using books and other reference materials in the library, look up the word quark. Find out what quarks are and how they were discovered. Describe some different types of quarks and what they do. Find out about the recent discoveries that have been made involving quarks. Report your findings to the class.

Becoming Stable

Imagine a large rock hanging over the edge of a cliff. How might you describe the rock's precarious position? You would probably say it is unstable, meaning that it cannot remain that way for long. Most likely, the rock will fall to the ground below, where it will be in a stable condition. Once it has fallen, the rock will certainly never move itself back to the cliff. Perhaps now you can think of an answer as to why an unstable nucleus breaks apart.

A nucleus that is unstable can become stable by undergoing a nuclear reaction, or change. There are four types of nuclear reactions that can occur. In each type, the identity of the original element is changed as a result of the reaction. You will now learn about each type of nuclear reaction.

Radioactive Decay

The process in which atomic nuclei emit particles or rays to become lighter and more stable is called **radioactive decay.** Radioactive decay is the spontaneous breakdown of an unstable atomic nucleus. There are three types of radioactive decay, each determined by the type of radiation released from the unstable nucleus.

ALPHA DECAY Alpha decay occurs when a nucleus releases an alpha particle. The release of an alpha particle (2 protons and 2 neutrons) decreases the mass number of the nucleus by 4. The mass number is the sum of the number of protons and neutrons in the nucleus. Each proton and each neutron has a mass of 1. The release of an alpha particle decreases the number of protons, or the atomic number, by 2. Thus the original atom is no longer the same. A new atom with an atomic number that is 2 less than the original is formed.

An example of an element that undergoes alpha decay is an isotope of uranium called uranium-238. The number 238 to the right of the hyphen is the mass number for this particular nucleus. An isotope of an element is often represented by using the element's symbol, mass number, and atomic number. The mass number is written to the upper left of the symbol. At the lower left, the atomic number (or number of protons) is written. Uranium has 92 protons. So this is the way uranium-238 would be represented:

$$^{238}_{92}\text{U}$$

The number of neutrons in the nucleus can be determined by subtracting the number of protons from the mass number. In this example, the number of neutrons is the mass number 238 minus the number of protons, 92, or 146 (238 − 92 = 146).

When uranium-238 undergoes alpha decay, or loses an alpha particle, it changes into an atom of thorium (Th), which has 90 protons and 144 neutrons. What is the mass number of thorium?

BETA DECAY   Beta decay occurs when a beta particle is released from a nucleus. As you have learned, a beta particle is an electron formed inside the nucleus when a neutron breaks apart. The other particle that forms when a neutron breaks apart is a proton. So beta decay produces a new atom with the same mass number as the original atom but with an atomic number one higher than the original atom. The atomic number is one higher because there is now an additional proton.

An example of an element that undergoes beta decay is carbon-14. An atom of carbon-14 has 6 protons and 8 neutrons. During beta decay it changes into an atom of nitrogen-14. An atom of nitrogen-14 has 7 protons and 7 neutrons.

When a nucleus releases either an alpha particle or a beta particle, the atomic number, and thus the identity, of the atom changes. **The process in which one element is changed into another as a result of changes in the nucleus is known as transmutation.** The word **transmutation** comes from the word *mutation*, which means change, and the prefix *trans-*, which means through.

GAMMA DECAY   Alpha and beta decay are almost always accompanied by gamma decay, which involves the release of a gamma ray. When a gamma ray is emitted by a nucleus, the nucleus does not change into a different nucleus. But because a gamma ray is an extremely high-energy wave, the nucleus makes a transition to a lower energy state.

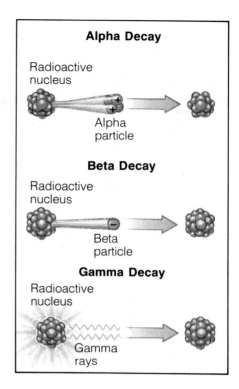

Figure 5–7 *Radioactive elements emit mass and energy during three different types of decay processes. During alpha decay, a helium nucleus and energy are released. A beta particle, or electron, and energy are released during beta decay. During gamma decay, high-energy electromagnetic waves are emitted. Which type of decay does not result in a different element?*

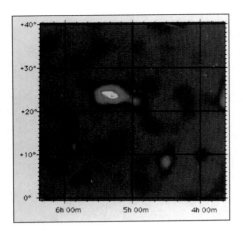

Figure 5–8 *Gamma rays are not as rare as you may think. For example, energy in the form of gamma rays is constantly being emitted from objects in space. This image of a solar flare was recorded by detecting gamma rays from the sun.*

Radioactive Half-Life

A sample of any radioactive element consists of a vast number of radioactive nuclei. These nuclei do not all decay at one time. Rather, they decay one by one over a period of time at a fixed rate. The fixed rate of decay of a radioactive element is called the **half-life.** The half-life is the amount of time it takes for half the atoms in a given sample of an element to decay.

The half-life of carbon-14 is 5730 years. In 5730 years, half the atoms in a given sample of carbon-14 will have decayed to another element: nitrogen-14. In yet another 5730 years, half the remaining carbon-14 will have decayed. At that time, one fourth—or one half of one half—of the original sample will be left. One fourth of the original sample will be carbon-14, and three fourths will be nitrogen-14.

Suppose you had 20 grams of pure barium-139. Its half-life is 86 minutes. So after 86 minutes, half the atoms in the sample would have decayed into another element: lanthanum-139. You would have 10 grams of barium-139 and 10 grams of lanthanum-139. After another 86 minutes, half the atoms in the 10 grams of barium-139 would have decayed into lanthanum-139. You would then have 5 grams of barium-139 and 15 grams of lanthanum-139. What would you have after the next half-life?

Figure 5–9 *The half-life of a radioactive element is the amount of time it takes for half the atoms in a given sample of the element to decay. After the first half-life, half the atoms in the sample are the radioactive element. The other half are the decay element, or the element into which the radioactive element changes. What remains after the second half-life? After the third?*

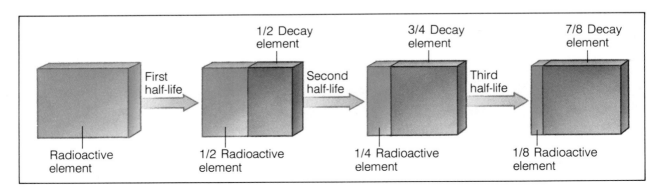

HALF-LIVES OF SOME RADIOACTIVE ELEMENTS

Element	Half-Life
Bismuth-212	60.5 minutes
Carbon-14	5730 years
Chlorine-36	400,000 years
Cobalt-60	5.26 years
Iodine-131	8.07 days
Phosphorus-32	14.3 days
Polonium-215	0.0018 second
Polonium-216	0.16 second
Radium-226	1600 years
Sodium-24	15 hours
Uranium-235	710 million years
Uranium-238	4.5 billion years

Figure 5–10 *The half-lives of radioactive elements vary greatly. Using the known half-lives of certain radioactive elements, such as carbon-14 and uranium-238, scientists can determine the age of ancient objects. Radioactive dating has been essential to the discovery of information about the Earth's history and about the evolution of organisms such as this ancient turtle.*

The half-lives of certain radioactive isotopes are useful in determining the ages of rocks and fossils. Scientists can use the half-life of carbon-14 to determine the approximate age of organisms and objects less than 50,000 years old. The technique is called carbon-14 dating. Other radioactive elements, such as uranium-238, can be used to date objects many millions of years old.

Half-lives vary greatly from element to element. Some half-lives are only seconds; others are billions of years. For example, the half-life of rhodium-106 is 30 seconds. The half-life of uranium-238 is 4.5 billion years!

Decay Series

As radioactive elements decay, they change into other elements. These elements may in turn decay, forming still other elements. The spontaneous breakdown continues until a stable, nonradioactive nucleus is formed. The series of steps by which a radioactive nucleus decays into a nonradioactive nucleus is called a **decay series.** Figure 5–11 on page 122 shows the decay series for uranium. What stable nucleus results from this decay series?

Because of the occurrence of decay series, certain radioactive elements are found in nature that otherwise would not be. In the 5-billion-year history of the solar system, many isotopes with short half-lives have decayed quickly. Thus they should not exist in

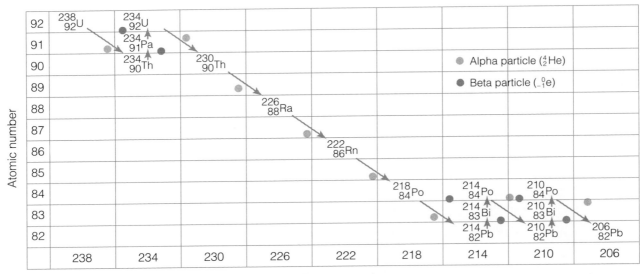

Figure 5–11 *The decay series for uranium-238 is shown in this graph. Radioactive uranium forms nonradioactive lead as a result of the decay series. What happens to the atomic mass number during a decay series? The atomic number?*

The Domino Effect, p.159

nature today. This is hardly the case, however. For example, radium, whose half-life is 1600 years, should have disappeared long ago. Yet it still exists on Earth today. This is because radium is part of the decay series for an isotope with a much longer half-life: uranium-238. Recall that uranium-238 has a half-life of 4.5 billion years.

Artificial Transmutation

Once scientists understood how natural transmutation occurred, they worked to produce **artificial transmutation.** The key was to find a way to change the number of protons in the nucleus of an atom. Ernest Rutherford, the same scientist who discovered the nucleus of the atom, produced the first artificial transmutation. By using alpha particles emitted during the radioactive decay of radium to bombard (hit forcefully) nitrogen nuclei, he produced an isotope of oxygen.

Getting the particles to hit the target nuclei with enough force to alter them is extremely difficult. In order to more effectively bombard nuclei with high-energy particles, scientists have developed devices for accelerating (speeding up) charged particles. One such device is the supercollider you read about at the beginning of this chapter. Other devices are the cyclotron, synchrotron, betatron, and linear accelerator. These devices use magnets and electric fields to speed up particles and produce collisions.

Before the discovery of the neutron in 1932, mainly alpha particles and protons were used as the "bullets" to bombard nuclei. But because both these particles are positively charged, they are repelled by the positive charge of the target nucleus. A large amount of extra energy is required simply to overcome this repulsion.

Enrico Fermi (FER-mee), an Italian scientist, and his co-workers realized that because neutrons are neutral, they are not repelled by the nucleus. These researchers discovered that neutrons can penetrate the nucleus of an atom more easily than a charged particle can. Neutrons can go through the nucleus without changing it; they can cause the nucleus to disintegrate; or they can become trapped by the nucleus, causing it to become unstable and break apart.

After a great deal of experimentation, the elements neptunium and plutonium were created. They were the first **transuranium elements.** Transuranium elements (also known as synthetic elements) are those with more than 92 protons in their nuclei. In other words, transuranium elements have atomic numbers greater than 92. A whole series of transuranium elements have been formed by bombarding atomic nuclei with neutrons, alpha particles, or other nuclear "bullets."

Figure 5–12 *Artificial transmutation of elements is done in a particle accelerator, such as the one at Fermilab in Illinois. This aerial view shows the outline of the underground tunnel. Particles traveling through long tubes will reach a final speed greater than 99.999 percent of the speed of light!*

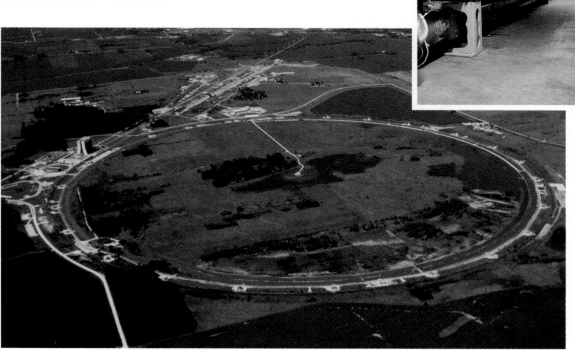

Figure 5–13 *Some elements present at the origin of the universe no longer exist. Others, whose half-lives are so short that they should no longer exist, do exist because they are part of the decay series of other radioactive elements. Scientists study decay series in an effort to learn more about the universe and its structures, such as the Tarantula Nebula.*

Radioactive isotopes of natural elements can be made by using a similar technique. Marie Curie's daughter, Irene, and Irene's husband, Frederic Joliot, discovered that stable atoms can be made radioactive when they are bombarded with neutrons. For example, by shooting neutrons at the nucleus of an iodine atom, scientists have been able to make I-131, a radioactive isotope of iodine.

ACTIVITY
THINKING

Interpreting Nuclear Equations

Nuclear reactions can be described by equations in much the same way chemical reactions can. Symbols are used to represent atoms as well as particles. And the total mass numbers on both sides of the equation are equal. The following reaction describes the alpha decay of uranium-238.

$$^{238}_{92}U \rightarrow\ ^{234}_{90}Th +\ ^{4}_{2}He + \text{Gamma rays}$$

Notice how an alpha particle is written as a helium nucleus. Other particles are written in a similar manner: a beta particle is written $^{0}_{-1}e$, and a neutron is written $^{1}_{0}n$.

Describe what is happening in each of the following:

$$^{235}_{92}U +\ ^{1}_{0}n \rightarrow\ ^{90}_{37}Rb +\ ^{144}_{55}Cs + 2\,^{1}_{0}n$$

$$^{210}_{82}Pb \rightarrow\ ^{210}_{83}Bi +\ ^{0}_{-1}e$$

$$^{222}_{86}Rn \rightarrow\ ^{4}_{2}He +\ ^{218}_{84}Po$$

5–2 Section Review

1. How is the binding energy related to the stability of a nucleus? How can an unstable nucleus become stable?
2. What happens during radioactive decay? What are three types of radioactive decay?
3. What is half-life? What is a decay series?
4. What is transmutation? Artificial transmutation?

Critical Thinking—*Making Calculations*
5. The half-life of radium-222 is 38 seconds. How many grams of radium-222 remain in a 12-gram sample after 76 seconds? After 114 seconds? How many half-lives have occurred when 0.75 gram remains?

CONNECTIONS

An Invisible Threat

When you think of the greatest threats to the environment and to your health and safety, scenes of polluted waterways and dirty landfills probably come to mind. So you might be surprised to learn that one of the most widespread and serious environmental threats cannot even be seen. This dangerous menace is radon.

Radon-222 is a radioactive element (atomic number 86) that is produced as part of the decay series for uranium-238. Because of uranium's long half-life, radon is continuously being generated. Since the mid-1980s, radon has become a priority concern for the United States Environmental Protection Agency (EPA). In addition to producing the effects normally associated with nuclear radiation, highly concentrated radon in the air can cause lung cancer if inhaled in large quantities. Radon represents one of the few naturally existing pollutants.

The EPA has set limits for radon levels in the home. However, many homes have radon levels nearly 1000 times greater than the limit. One reason for this is that many buildings are constructed on land rich in uranium ore. Radon is released from soil and rocks containing this ore. In the outdoors, radon is diluted to safe levels. But when it leaks into buildings through cracks in basement floors and walls, radon is trapped—and dangerous.

The concentration of radon in a building depends on the type of construction

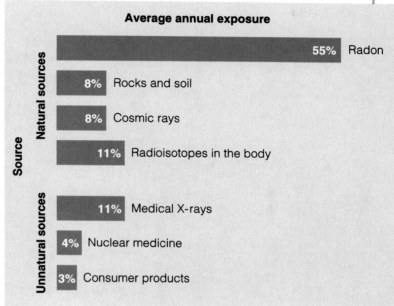

Average annual exposure

and the materials used. New energy-efficient buildings, which are designed to keep heated or cooled air in, also trap radon. One way to protect against radon, especially in areas with a high natural concentration, is to seal cracks in foundation walls and floors. Another way is to improve air ventilation by circulating outside air into a building, thereby diluting existing radon concentrations.

Over the long term, everyone is exposed to potentially damaging radiation from both natural sources and human activity. But radon exposure currently accounts for more than half the average annual exposure to radiation. Home radon-testing kits have become a common household item in many parts of the country. One of the most important protections against the health hazards of radon is the awareness that something that cannot be seen, touched, or smelled can be life-threatening.

Guide for Reading

Focus on this question as you read.

▶ What is the difference between nuclear fission and nuclear fusion?

5-3 Harnessing the Nucleus

Radioactive decay and the bombardment of a nucleus with particles are two ways in which energy is released from the nucleus of an atom. The amount of energy released, however, is small compared with the tremendous amount of energy known to bind the nucleus together. Long ago, scientists realized that if somehow they could release more of the energy holding the nucleus together, huge amounts of energy could be gathered from tiny amounts of mass.

Nuclear Fission

During the 1930s, several other scientists built upon the discovery of Fermi and his co-workers. In 1938, the German scientists Otto Hahn and Fritz Strassman discovered that when the nucleus of an atom of uranium-235 is struck by a neutron, two smaller nuclei of roughly equal mass were produced. Two other scientists, Lise Meitner and Otto Frisch, provided an explanation for this event: The uranium nucleus had actually split into two. What made the discovery and the explanation so startling was that until then the known nuclear reactions had involved only knocking out a tiny fragment from the nucleus—not splitting it into two!

This reaction—the first of its kind ever to be produced—is an example of **nuclear fission** (FIHSH-uhn). It was so named because of its resemblance to cell division, or biological fission. **Nuclear fission is the splitting of an atomic nucleus into two smaller nuclei of approximately equal mass.** Unlike radioactive decay, nuclear fission does not occur spontaneously.

In one typical fission reaction, a uranium-235 nucleus is bombarded by a neutron, or nuclear "bullet." The products of the reaction are a barium-141 nucleus and a krypton-92 nucleus. Three neutrons are also released: the original "bullet" neutron and 2 neutrons from the uranium nucleus.

The amount of energy released when a single uranium-235 nucleus splits is not very great. But the neutrons released in the first fission reaction become nuclear "bullets" that are capable of splitting other

Figure 5–14 *In the midst of the social and political strife of the late 1930s, scientists such as Lise Meitner and Otto Hahn succeeded in recognizing and explaining the events of nuclear fission. What is nuclear fission?*

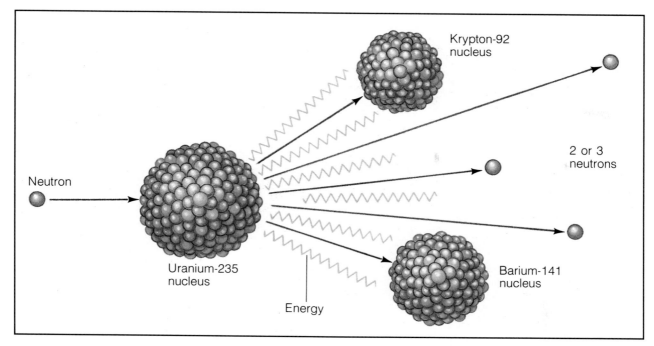

Figure 5–15 *In this diagram, a uranium-235 nucleus is bombarded with a neutron. The nucleus breaks up, producing a nucleus of krypton-92 and a nucleus of barium-141. Large amounts of energy as well as two additional neutrons are released. Each neutron is capable of splitting another uranium-235 nucleus. What is this repeating process called?*

uranium-235 nuclei. Each uranium nucleus that is split releases 3 neutrons. These neutrons may then split even more uranium nuclei. The continuous series of fission reactions is called a nuclear chain reaction. In a **nuclear chain reaction,** billions of fission reactions may take place each second!

When many atomic nuclei are split in a chain reaction, huge quantities of energy are released. This energy is produced as a result of the conversion of a small amount of mass into a huge amount of energy. The total mass of the barium, krypton, and 2 neutrons is slightly less than the total mass of the original uranium plus the initial neutron. The missing mass has been converted into energy. An uncontrolled chain reaction produces a nuclear explosion. The atomic bomb is an example of an uncontrolled chain reaction.

All currently operating nuclear power plants use controlled fission reactions to produce energy. The energy is primarily in the form of heat. The heat is carried away and used to produce electricity.

ACTIVITY READING

An Event That Would Change History Forever

With the excitement of science and the versatility of technology come serious responsibilities and social concerns. The development and use of the atomic bomb was one such ramification of the discovery of nuclear fission. Read *Hiroshima* by John Hersey to gain some insight into the effects such an event has had on people, politics, and history.

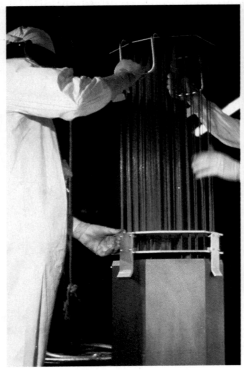

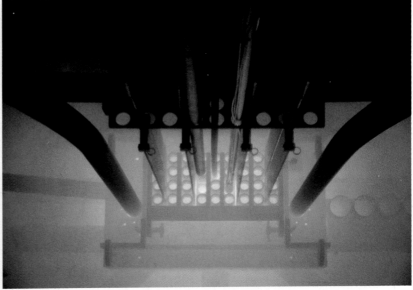

Figure 5–16 *Inside a nuclear reactor, radioactive fuel rods give off energy that produces a blue glow in the water. Before being placed in the reactor, a bundle of uranium-containing rods must be carefully checked.*

Nuclear Fusion

Another type of nuclear reaction that certain radioactive elements can undergo is called **nuclear fusion** (FYOO-zhuhn). Like fission, this kind of nuclear reaction produces a great amount of energy. But unlike fission, which involves the splitting of a high-mass nucleus, this reaction involves the joining of two low-mass nuclei. The word fusion means joining together. **Nuclear fusion is the joining of two atomic nuclei of smaller masses to form a single nucleus of larger mass.**

Nuclear fusion is a thermonuclear reaction. The prefix *thermo-* means heat. For nuclear fusion to take place, temperatures well over a million degrees Celsius must be reached. At such temperatures, the phase of matter known as plasma is formed. Plasma consists of positively charged ions, which are the nuclei of original atoms, and free electrons.

The temperature conditions required for nuclear fusion exist on the sun and on other stars. In fact, it is nuclear fusion that produces the sun's energy. In the sun's core, temperatures of about 20 million degrees Celsius keep fusion going continuously. In a series of steps, hydrogen nuclei are fused into a helium-4 nucleus. See Figure 5–18.

Nuclear fusion produces a tremendous amount of energy. The energy comes from matter that is converted into energy during the reaction. In fact,

the products formed by fusion have a mass that is about 1 percent less than the mass of the reactants. Although 1 percent loss of mass may seem a small amount, its conversion produces an enormous quantity of energy.

Nuclear fusion has several advantages over nuclear fission. The energy released in fusion reactions is greater for a given mass than that in fission reactions. Fusion reactions also produce less radioactive waste. And the possible fuels used for fusion reactions are more plentiful. Unfortunately, considerable difficulties exist with producing useful

Figure 5–17 *The light and heat that make life on Earth possible are the result of nuclear fusion within the sun. The destruction caused by a hydrogen bomb is also the result of nuclear fusion. The only source capable of delivering the energy required to trigger a fusion reaction in the hydrogen bomb is an atomic explosion.*

Figure 5–18 *In the process of nuclear fusion, hydrogen nuclei fuse to produce helium and tremendous amounts of energy. What other products are formed during the fusion reaction?*

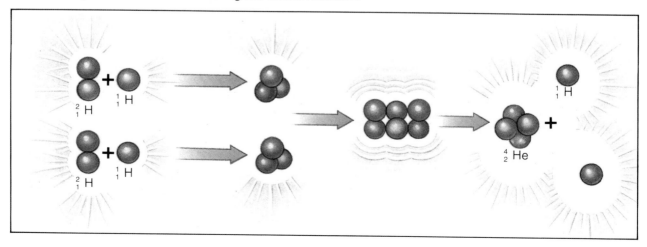

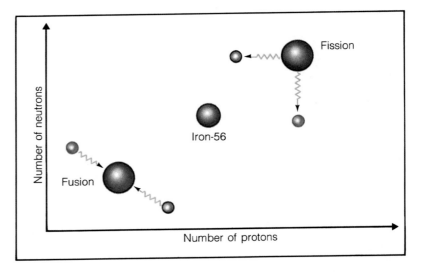

Figure 5–19 *Different elements will give off energy and become more stable by undergoing either fusion or fission. For most elements lighter than iron-56, nuclear fusion will give off energy. For most elements heavier than iron-56, nuclear fission will give off energy. Elements gain stability by moving closer on the graph to iron-56.*

fusion reactions on Earth. Fusion reactions are more difficult to begin, to control, and to maintain than nuclear fission reactions are. After all, no known vessel can contain reactions occurring at such tremendous temperatures. And such high temperatures are extremely difficult to achieve. In fact, a hydrogen bomb, which uses fusion, is started by an atomic bomb, which uses fission. It is the only way of achieving the necessary temperatures.

Scientists are continuing their search for ways to control this powerful reaction and to tap a tremendous energy resource. As an example, experiments using high-powered laser beams and electrons as ways of starting fusion reactions are being conducted.

5–3 Section Review

1. What is nuclear fission? Nuclear fusion?
2. How is the sun's energy produced?
3. Where does the energy produced in both fission and fusion reactions come from?
4. Compare the energy produced by fission and fusion reactions with the energy produced by radioactive decay.

Connection—*You and Your World*
5. Why is the production of less radioactive waste considered an advantage of nuclear fusion?

5–4 Detecting and Using Radioactivity

Guide for Reading

Focus on these questions as you read.

▶ *What instruments are used to detect and measure radioactivity?*

▶ *What are some uses of radioactive substances?*

Radioactivity cannot be seen or felt. Becquerel discovered radioactivity because it left marks on photographic film. Although film is still used today to detect radioactivity, scientists have more specialized instruments for this purpose. **The instruments scientists use to detect and measure radioactivity include the electroscope, the Geiger counter, the cloud chamber, and the bubble chamber.**

Instruments for Detecting and Measuring Radioactivity

ELECTROSCOPE An **electroscope** is a simple device that consists of a metal rod with two thin metal leaves at one end. If an electroscope is given a negative charge, the metal leaves separate. In this condition, the electroscope can be used to detect radioactivity.

Radioactive substances remove electrons from molecules of air. As a result, the molecules of air become positively charged ions. When a radioactive substance is brought near a negatively charged electroscope, the air molecules that have become positively charged attract the negative charge on the leaves of the electroscope. The leaves discharge, or lose their charge, and collapse.

GEIGER COUNTER In 1928, Hans Geiger designed an instrument that detects and measures radioactivity. Named the **Geiger counter** in honor of its inventor, this instrument produces an electric current in the presence of a radioactive substance.

A Geiger counter consists of a tube filled with a gas such as argon or helium at a reduced pressure. When radiation enters the tube through a thin window at one end, it removes electrons from the atoms of the gas. The gas atoms become positively charged ions. The electrons move through the positively charged ions to a wire in the tube, setting up an electric current. The current, which is amplified and fed into a recording or counting device, produces a flashing light and a clicking sound. The number of

Figure 5–20 *Because radioactive substances will cause an electroscope to discharge, an electroscope can be used to detect radiation. What do radioactive substances do to molecules of air?*

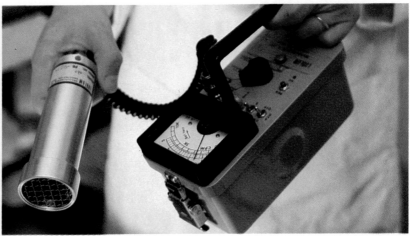

Figure 5–21 *A Geiger counter detects and measures radioactivity. A spiderwort plant is nature's radiation detector. The stamens of the spiderwort flower are usually blue. In the presence of radiation the stamens turn pink.*

flashes and clicks per unit time indicates the strength of the radiation. A counter attached to the wire is able to measure the amount of radioactivity by measuring the amount of current.

CLOUD CHAMBER A **cloud chamber** contains a gas cooled to a temperature below its usual condensation point (point at which it becomes a liquid). When a radioactive substance is put inside the chamber, droplets of the gas condense around the radioactive particles. The process is similar to what happens in "cloud seeding," when rain droplets condense around particles that have been injected into the clouds. The droplets formed around the particles of radiation in a cloud chamber leave a trail that shows up along the chamber lining. An alpha particle leaves a short, fat trail, whereas a beta particle's trail is long and thin.

BUBBLE CHAMBER The **bubble chamber** is similar in some ways to the cloud chamber, although its construction is more complex. A bubble chamber contains a superheated liquid. A superheated liquid is hot enough to boil—but does not. Instead, it remains in the liquid phase. The superheated liquid most often contained in a bubble chamber is hydrogen.

When radioactive particles pass through the chamber, they cause the hydrogen to boil. The boiling liquid leaves a trail of bubbles, which is used to track the radioactive particle. The photograph you saw at the beginning of this chapter is of a bubble chamber after a nuclear reaction.

Figure 5–22 *This cloud chamber photograph shows an upward stream of alpha particles.*

Putting Radioactivity to Work

Radioactive substances have many practical uses. Dating organic objects, which you learned about earlier, is one such use. In industry, radioactive isotopes, or **radioisotopes,** have additional uses. Radioisotopes can be used to find leaks or weak spots in metal pipes, such as oil pipe lines. Radioisotopes also help study the rate of wear on surfaces that rub together. One surface is made radioactive. Then the amount of radiation on the other surface indicates the wear.

Because radioisotopes can be detected so readily, they can be used to follow an element through an organism, or through an industrial process, or through the steps of a chemical reaction. Such a radioactive element is called a **tracer,** or radiotracer. Tracers are possible because all isotopes of the same element have essentially the same chemical properties. When a small quantity of radioisotope is mixed with the naturally occurring stable isotopes of the same element, all the isotopes go through the same reactions together.

An example of a tracer is phosphorus-32. The nonradioactive element phosphorus is used in small amounts by both plants and animals. If phosphorus-32 is given to an organism, the organism will use the radioactive phosphorus just as it does the nonradioactive phosphorus. However, the path of the radioactive element can be traced. In this way, scientists can learn a great deal about how plants and animals use phosphorus.

Another area in which radioisotopes make an important contribution is in the field of medicine. The branch of medicine in which radioactivity is used is known as nuclear medicine. Tracers are extremely valuable in diagnosing diseases. For example, radioactive iodine—iodine-131—can be used to study the function of the thyroid gland, which absorbs iodine. Sodium-24 can be used to detect diseases of the circulatory system. Iron-59 can be used to study blood circulation.

Another procedure, known as radioimmunoassay, developed by Dr. Rosalyn Yalow—who won the Nobel prize for her work—involves using tracers to detect the presence of minute quantities of substances in

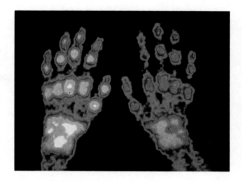

Figure 5–23 *Radioisotopes can be used as tracer elements to produce images such as this one of a person's hands. The different-colored areas help doctors diagnose conditions in order to prescribe treatment. This person has arthritis.*

ACTIVITY

DOING

Radioactivity in Medicine

Obtain permission to visit the radiation laboratory at a local hospital. Observe the instruments used and the safety precautions taken for both patients and technicians when using the devices. Interview a technician, if possible. Make a report on your visit.

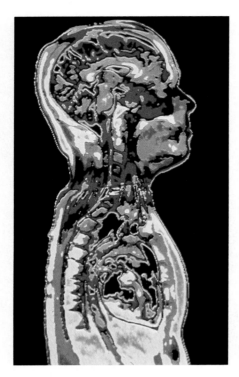

Figure 5–24 *Observing changes in the nuclei of atoms has many important applications. Through nuclear magnetic resonance imaging, doctors can form detailed images of various parts of the human body. What is an advantage of this procedure over exploratory surgery?*

the body. These tests can be used to detect pregnancy as well as the early signs of a disease. Another powerful research tool—nuclear magnetic resonance imaging (MRI)—has become invaluable in a variety of fields, from physics to chemistry and biochemistry. MRI involves recording changes in the energy of atomic nuclei in response to external energy changes, without altering the cells of the body in any way.

Radiation is also used to destroy unhealthy cells, such as those that cause cancer. Radiation in large doses destroys living tissues, especially cells undergoing division. Because cancer cells undergo division more frequently than normal cells do, radiation kills more cancer cells than it does normal cells. As early as 1904, physicians attempted to treat masses of unhealthy cells, known as tumors, with high-energy radiation. This treatment is called radiation therapy.

Radioisotopes can also be used to kill bacteria that cause food to spoil. Radiation was used to preserve the food that the astronauts ate while on the moon and in orbit.

Dangers of Radiation

Although radioactivity has tremendous positive potential, radioactive materials must be handled with great care. Radioactive materials are extremely dangerous. Radiation can ionize—or knock electrons out of—atoms or molecules of any material it passes through. For this reason, the term ionizing radiation is sometimes used. So, oddly enough, the same radiation that is used to treat disease can also cause it.

Ionization can cause considerable damage to materials, particularly to biological tissue. When ionization is produced in cells, ions may take part in chemical reactions that would not otherwise have occurred. This may interfere with the normal operation of the cell. Damage to DNA is particularly serious. An alteration in the DNA (substance responsible for carrying traits from one generation to another) of a cell can interfere with the production of proteins and other essential cellular materials. The result may be the death of the cell. If many cells die, the organism may not be able to survive.

Large doses of radiation can cause reddening of the skin, a drop in the white blood cell count, and numerous other unpleasant symptoms, including nausea, fatigue, and loss of hair. Such effects are sometimes referred to as radiation sickness. Large doses of radiation can also be fatal. Marie Curie's death in 1934 was caused by exposure to too much radiation.

Even metals and other structural materials can be weakened by intense radiation. This is a considerable problem in nuclear-reactor power plants and for space vehicles that must pass through areas of intense cosmic radiation.

We are constantly exposed to low-level radiation from natural sources such as cosmic rays from space, radioactivity in rocks and soil, and radioactive isotopes that are present in food and in our bodies.

Today, people who work with radioactive materials take extreme precautions. They wear radiation-sensitive badges that serve as a warning of unsafe levels of radiation. Specially designed clothing is worn to block radiation. Scientists continue to search for greater understanding and control of radiation so that its benefits can be enjoyed without the threat of danger.

Figure 5–25 *The hands and numbers of a luminous watch contain minute amounts of radium. The radioactive decay of radium causes the watch to glow in the dark. People who painted the dials on clocks in the early 1900s suffered from radiation poisoning because they often licked the tips of their brushes to make fine lines.*

Figure 5-26 *Workers handling radioactive materials must use extreme caution and wear specially designed clothes for protection.*

5–4 Section Review

1. Name four instruments used to detect and measure radioactivity.
2. Compare a bubble chamber and a cloud chamber.
3. What is a radioisotope? What are its uses?
4. What is a tracer? Describe several uses of tracers.

Connection—*Life Science*
5. A mutation is a change that occurs in the genetic material of a cell (the code that determines what traits will be carried on from generation to generation). Why does nuclear radiation cause mutations?

Laboratory Investigation

The Half-Life of a Sugar Cube

Problem

How can the half-life of a large sample of sugar cubes be determined?

Materials *(per group)*

250 sugar cubes	large bowl
food coloring	medicine dropper

Procedure 🔬

1. Place a small drop of food coloring on one side of each sugar cube.

2. Put all the sugar cubes in a bowl. Then gently spill them out on the table. Move any cubes that are on top of other cubes.

3. Remove all the sugar cubes that have the colored side facing up. If you have room on the table, arrange in a vertical column the sugar cubes that you removed. Put the rest of the cubes back in the bowl.

4. Repeat step 3 several more times until five or fewer sugar cubes remain.

5. On a chart similar to the one shown, record the number of tosses (times you spilled the sugar cubes), the number of sugar cubes removed each time, and the number of sugar cubes remaining. For example, suppose after the first toss you removed 40 sugar cubes. The number of tosses would be 1, the number of cubes removed would be 40, and the number of cubes remaining would be 210 (250 – 40).

Observations

1. Make a full-page graph of tosses versus cubes remaining. Place the number of tosses on the X (horizontal) axis and the number of cubes remaining on the Y (vertical) axis. Start at zero tosses with all 250 cubes remaining.

2. Determine the half-life of the decaying sugar cubes in the following way. Find the point on the graph that corresponds to one half of the original sugar cubes (125). Move vertically down from this point until you reach the horizontal axis. Your answer will be the number of tosses.

Tosses	Sugar Cubes Removed	Sugar Cubes Remaining
0	0	250
1	40	210
2		
3		

Analysis and Conclusions

1. What is the shape of your graph?

2. How many tosses are required to remove one half of the sugar cubes?

3. How many tosses are required to remove one fourth of the sugar cubes?

4. Assuming tosses are equal to years, what is the half-life of the sugar cubes?

5. Using your answer to question 4, how many sugar cubes should remain after 8 years? After 12 years? Do these numbers agree with your observations?

6. What factor(s) could account for the differences in your observed results and those calculated?

7. **On Your Own** Repeat the experiment with a larger number of sugar cubes. Predict whether the determined half-life will be different. Is it?

Summarizing Key Concepts

5–1 Radioactivity

▲ An element that gives off nuclear radiation is said to be radioactive.

▲ Nuclear radiation occurs in three forms: alpha particles, beta particles, and gamma rays.

5–2 Nuclear Reactions

▲ If the binding energy—the force that holds the nucleus together—is not strong, an atom is said to be unstable. Atoms with unstable nuclei are radioactive.

▲ Atoms that have the same atomic number but different numbers of neutrons are called isotopes.

▲ An unstable nucleus eventually becomes stable by undergoing a nuclear reaction.

▲ Radioactive decay is a nuclear reaction that involves the spontaneous breakdown of an unstable nucleus. During radioactive decay, alpha particles, beta particles, and/or gamma rays are emitted.

▲ The decay of a radioactive element occurs at a fixed rate called the half-life.

▲ The series of steps by which a radioactive nucleus decays is known as a decay series.

▲ Artificial transmutation involves the nuclear reactions in which atomic nuclei are bombarded with high-speed particles.

5–3 Harnessing the Nucleus

▲ Nuclear fission is the splitting of an atomic nucleus to form two smaller nuclei of roughly equal mass.

▲ Nuclear fusion is the joining together of two atomic nuclei to form a single nucleus of larger mass.

5–4 Detecting and Using Radioactivity

▲ Four devices that can detect radioactivity are the electroscope, Geiger counter, cloud chamber, and bubble chamber. The Geiger counter can also measure radioactivity.

▲ Radioactive substances must be handled carefully because large amounts of radiation can be harmful to living things.

Reviewing Key Terms

Define each term in a complete sentence.

5–1 Radioactivity
nuclear radiation
radioactive
radioactivity
alpha particle
beta particle
gamma ray

5–2 Nuclear Reactions
nuclear strong force
binding energy

isotope
radioactive decay
transmutation
half-life
decay series
artificial transmutation
transuranium element

5–3 Harnessing the Nucleus
nuclear fission

nuclear chain reaction
nuclear fusion

5–4 Detecting and Using Radioactivity
electroscope
Geiger counter
cloud chamber
bubble chamber
radioisotope
tracer

Chapter Review

Content Review

Multiple Choice

Choose the letter of the answer that best completes each statement.

1. The particle given off by a radioactive element that is actually a helium nucleus is a(an)
   a. beta particle.   c. isotope.
   b. gamma ray.   d. alpha particle.
2. Atoms with the same atomic number but different numbers of neutrons are
   a. isotopes.   c. alpha particles.
   b. radioactive.   d. beta particles.
3. In relation to the original atom, the atom that results from alpha decay has an atomic number that is
   a. 2 less.   c. the same.
   b. 1 less.   d. 2 more.
4. The process in which one element changes into another as a result of nuclear changes is
   a. fluorescence.   c. transuranium.
   b. transmutation.   d. synthesis.

5. An atomic nucleus splits into two smaller nuclei in
   a. fusion.   c. fission.
   b. alpha decay.   d. transmutation.
6. A device in which radioactive materials leave a trail of liquid droplets is a(an)
   a. bubble chamber.   c. decay chamber.
   b. cloud chamber.   d. electroscope.
7. A Geiger counter detects radioactivity when the radioactive substance
   a. leaves a trail of bubbles.
   b. condenses around particles of a gas.
   c. causes liquid gas to boil.
   d. produces an electric current.
8. An artificially produced radioactive isotope of an element is called a
   a. synthetic isotope.   c. radioisotope.
   b. transmutation.   d. gamma isotope.

True or False

If the statement is true, write "true." If it is false, change the underlined word or words to make the statement true.

1. One of the radioactive elements discovered by the Curies was <u>uranium</u>.
2. The energy that holds the nucleus together is the <u>binding energy</u>.
3. <u>Gamma rays</u> are electromagnetic waves of very high frequency and energy.
4. The spontaneous breakdown of an unstable nucleus is <u>artificial transmutation</u>.
5. Two atomic nuclei join together during nuclear <u>fission</u>.
6. Gas molecules are ionized by a radioactive substance in a <u>bubble chamber</u>.
7. <u>Iodine-131</u> can be used to study the function of the thyroid gland.
8. <u>Radioisotopes</u> are used to kill bacteria.

Concept Mapping

Complete the following concept map for Section 5–1. Refer to pages O6–O7 to help you construct a concept map for the entire chapter.

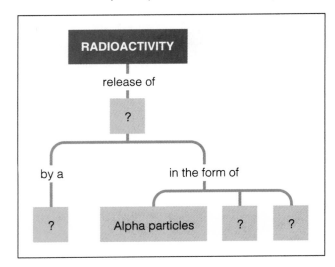

Concept Mastery

Discuss each of the following in a brief paragraph.

1. Describe how Becquerel's work illustrates the scientific method.
2. Describe the penetrating power of each of the three types of radiation.
3. How was the existence of the nuclear strong force deduced? What is the energy associated with the strong force, and how is it related to the stability of an element?
4. What do isotopes of a given element have in common? How are they different?
5. Describe the four types of nuclear reactions.
6. Describe how a Geiger counter works. How is it different from a bubble chamber, cloud chamber, and an electroscope?
7. Explain why the amount of helium in the sun is increasing.
8. Why must the fission process release neutrons if it is to be useful?
9. What are some of the uses of radioactivity?

Critical Thinking and Problem Solving

Use the skills you have developed in this chapter to answer each of the following.

1. **Making graphs** Sodium-24 has a half-life of 15 hours. Make a graph to show what happens to a 100-gram sample of sodium-24 over a 5-day period.
2. **Interpreting a graph** The dots on the accompanying graph represent stable nuclei. The straight line represents equal numbers of protons and neutrons. Describe the relationship between the number of protons and the number of neutrons as the atomic number increases.

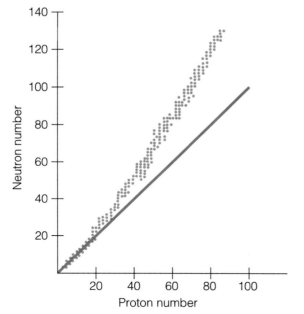

3. **Making comparisons** Describe the three types of radiation and relate them to the three types of radioactive decay.
4. **Analyzing data** A skeleton of an ancient fish is found to contain one eighth the amount of carbon-14 that it contained when it was alive. How old is the skeleton?
5. **Making calculations** The half-life of cobalt-60 is 5.26 years. How many grams of a 20-gram sample of cobalt-60 remain after 10.52 years? After 15.78 years?
6. **Drawing conclusions** Why do you think it is more dangerous for a woman of child-bearing age to be exposed to nuclear radiation than it is for a younger or an older woman to be exposed?
7. **Using the writing process** Food spoils because organisms such as bacteria and mold grow on it and decompose it. Gamma rays can destroy such organisms. So foods treated with gamma rays, or irradiated foods, last longer. Many people, however, are afraid that irradiated foods may be dangerous. Others do not want to live near gamma-ray treatment centers. Write a letter to your state representative describing your thoughts about irradiated foods.

ORGANIC BALLS:
New Adventures In Chemistry

You can't kick it through goalposts, dribble it down a basketball court, or hit it with a bat or a tennis racket. But scientists are extremely excited about a ball that has much more to do with atoms than with athletics!

The special sphere is called the "bucky ball." It is a molecular form of pure carbon shaped like the geodesic domes designed by the famous architect R. Buckminster Fuller. Thus its nickname—bucky ball—and its complete name: Buckminsterfullerene.

The humorous scientists who named the fullerene—as well as discovered it—include teams of chemists led by Richard Smalley of Rice University in Texas and Harry Kroto of the University of Sussex in England. In 1985, while researching the clustering properties of carbon atoms, the chemists noticed what they considered to be a curious trend: All the carbon molecules they produced contained an even number of atoms. Their measurements also indicated that a large proportion of the molecules consisted of 60 atoms. After further investigation—with equipment that ranged from sophisticated lasers to simple scraps of paper—the scientists theorized the existence of a special crystal form of carbon. Under the right circumstances, they claimed, 60 carbon atoms could bond together to form a hollow perfect sphere: the bucky ball. The sphere has 32 facets, or sides, of which 12 are pentagons (5-sided) and 20 are hexagons (6-sided). Its structure makes it an incredibly stable molecule.

Smalley and Kroto were offering the world a startling discovery: a third molecular form of pure carbon. The only other known forms of carbon are graphite and diamond. In light of its implications, Smalley and Kroto's proposal was first met with skepticism from

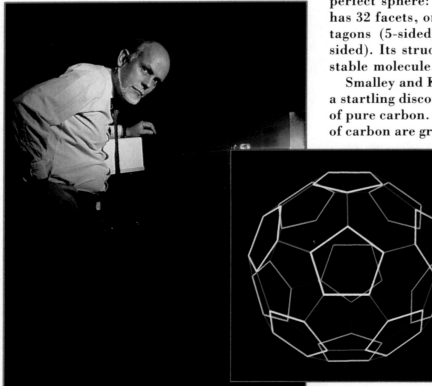

◀ **Dr. Richard Smalley is shown here at work in his laboratory blasting bits of carbon with a laser. His research, along with that of Dr. Kroto, resulted in the discovery of the bucky ball—seen here as a computer-generated image.**

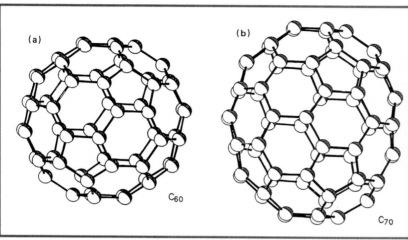

▲ ▶ A three-dimensional model of Buckminsterfullerene shows its pattern. Scientists hope that they can produce molecules that keep the same shape yet contain more carbon atoms. On the right is the first microscopic image of buckminsterfullerenes.

fellow scientists. Some said that Smalley and Kroto had failed to provide enough concrete evidence to prove their discovery. Some critics even said that the theory was too good, too simple, and too elegant to be true. But Smalley and Kroto refused to abandon their ideas. They were convinced that the Buckminsterfullerene molecule did indeed exist.

While Smalley and Kroto continued their research, other scientists around the globe began to investigate the possible existence of the bucky ball. In the fall of 1990, a good five years after Smalley and Kroto first published their findings, two major discoveries confirmed their theory. Researchers at the University of Arizona, the Max Planck Institute in Germany, and the International Business Machines Almaden Research Center in San Jose, California, produced and published the first photographs of the carbon balls. In addition, two physicists found a way to mass-produce the ball-shaped molecules. The laboratories in Arizona and Germany pioneered a method of producing the molecules by passing an electric arc between graphite electrodes. Carbon from the electrodes, which are kept in a vacuum with a small amount of helium, vaporizes and then condenses as soot. About 5 percent of this soot results in fullerene molecules.

Now scientists—even the skeptics—are excited about the fullerene's future. Because of its exceptional stability, the ball-shaped molecule has many potential applications. As a researcher at the University of Arizona explained, the fullerene may introduce a "brand new kind of chemistry." Scientists imagine that fullerenes may be able to make a new type of lubricant because the molecules roll like invisible ball bearings. The bucky balls may also be used to produce new organic compounds and in the development of better batteries. Other areas of research include the use of bucky balls on new telecommunications systems as building blocks for a new generation of high-speed computers based on light waves rather than on electricity. There is also hope of encapsulating drugs for cancer treatments in bucky balls to preserve the drugs until they reach the intended therapy site. But the busiest and most exciting area of research is in the application of fullerenes as superconductors.

So despite the fact that the bucky ball doesn't have much use on the playing fields, it undoubtedly has tremendous potential in the laboratory, as well as in the marketplace.

✳ HOW PRACTICAL IS ✳ FLOWER POWER?

Scientists are hard at work trying to find an energy substitute for oil, and it may be an ordinary green plant. But will the price be too high?

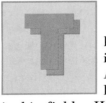

The sun rises on a summer day in the not-too-distant future. As its rays spread over the land, a farmer goes to work in his fields. His crop is almost ready for harvesting and will bring him a good price. What kind of crop is he growing? The farmer of the future is growing energy!

This may not seem like news to you. After all, some plants, such as trees, are already sources of energy. Trees are used as firewood. And when wood is burned, energy is released in the form of heat and light. Where does this energy come from? It comes from the sun!

Each green leaf in a plant uses energy from the sun—sunlight—to build chemical molecules of sugar. In the process, the sun's energy gets locked up in the sugar molecules and other plant substances. That energy is given off when a plant or its parts are burned.

The same thing happens when coal, oil, and natural gas are burned. These fuels were formed from decaying plants and from animals that ate plants. Buried in the Earth for millions of years, the remains of the plants and animals were converted by tremendous heat and pressure into the coal, oil, and gas of today. But only a small percentage of the sun's energy originally absorbed by the plants stayed in these fuels. So the burning of fuels is not the most efficient way to get at plant energy.

FUEL FROM PLANTS

Melvin Calvin, a Nobel prize-winning chemist at the University of California at Berkeley, has worked for years to find a better way to recover the solar energy stored in plants. One way to get more energy out of plants is to take their chemicals and make them into high-energy fuels. One such fuel is alcohol. When burned, alcohol is a powerful fuel. Some high-powered race cars, for instance, run on pure alcohol.

Unfortunately, alcohol is not made directly by plants but from their sugar compounds.

▲ Substitutes for oil and gas are always being hunted. Corn, for example, is a source of alcohol, which is combined with gasoline to make the fuel gasohol. But growing and harvesting corn uses up a lot of fuel and energy—too much, maybe, to make gasohol a practical substitute for gasoline.

Sugar cane and cereal grains such as corn produce large amounts of sugars. Through a chemical process called fermentation, these sugars can be broken down into alcohol.

A mixture of gasoline and alcohol produces gasohol, which is sold as a substitute for gasoline at some filling stations. When gasohol was first marketed, it was hailed as a solution to the energy crisis. But gasohol has never really become popular or practical. The following reasons highlight why "flower power" may not become a reality in the near future.

ENERGY TO MAKE ENERGY

For one thing, a lot of energy is needed to make energy. Corn, for example, is a major source of gasohol. Yet, one scientist notes, "Corn grown in the U.S. requires a surprising amount of energy."

For example, corn plants need large amounts of fertilizer. And most of the fertilizers used for growing corn are made from

▲ **Melvin Calvin stands in a field of gopher plants at his ranch in California. Calvin and other scientists like him hope to use gopher plants, and other plants, as new sources of fuel. Will gopher plants help to replace oil wells? Some scientists think so.**

oil. And, of course, it takes a lot of energy to get oil out of the ground and to move it to where it is used. Other ways that energy is used to produce corn are the burning of tractor fuel for planting, cultivating, and harvesting; the production of heat for grain drying; and, if irrigation is needed, energy to pump water.

Most of this energy comes from oil or natural gas. Moreover, converting the grain to alcohol requires even more energy. Equally important, if land is used to produce fuel, is how the world's food supplies would be affected. In other words, does it make sense to "farm" fuel? It makes sense only if no one goes without food and the fuel is inexpensive. So far, this does not seem to be the case.

However, scientists have come up with a possible way to get around the problem—

farm plants that *directly* produce fuel. Oil from plants sounds impossible, doesn't it? Some plants, however, produce hydrocarbons, which are the basic compounds in oil. Among these plants are the rubber tree and its relatives. Some of these plants resemble cacti. The hydrocarbons are in the milky sap of the plants. They might be used to produce inexpensive fuels. Melvin Calvin and other researchers have been trying to find a way to produce fuel from plants.

Calvin has also done research on the gopher plant, which grows in the western part of the United States. He and scientists working with him are growing gopher plants on experimental plots on Calvin's own ranch in California. They hope to determine the amount of hydrocarbons the gopher plants produce and the cost of making fuel from them.

Because gopher plants are wild and have not been cultivated before, little is known about methods of growing them and the conditions they require. The sap of the gopher plant, moreover, cannot be tapped as easily as that of some of its relatives.

Perhaps the gopher plant will prove to be too difficult to grow. Or possibly the amount of hydrocarbons that can be produced from it will not be large enough to make it a good fuel source. On the other hand, the gopher plant could turn out to be a substitute for an oil well. And if it is not, perhaps some other plant will be.

There are, of course, other sources of inexpensive energy we should look to: solar energy, nuclear energy, natural gas energy, old-fashioned coal, oil, and wind energy. But each has its problems. Can "flower power" compete with these energy sources? What do you think?

SCIENCE GAZETTE

The Right Stuff— PLASTICS

Cars, buildings, and appliances of the twenty-first century will be lighter, safer, and stronger than anything we have today.

Standing apart from a group of her classmates, Jennifer bounced her glass ball on the sidewalk. She was not paying attention to the other members of her class or to what her teacher, Ms. Parker, was saying.

As the sparkling ball bounced, Jennifer was barely aware of the sidewalk changing color to reduce the glare of sunlight. She did not notice a young woman hurrying along the street carrying an auto engine in one hand and groceries in the other. She did not look up as cars sped by towing boats and houses by plastic threads.

Jennifer and her Design for Living class were on a field trip to a brand-new "house of the future." The house had just been completed that year: 2093. It certainly made the students' homes seem old-fashioned.

For days before the field trip, Ms. Parker had talked about the house of the future. She had described how safety, security, cleaning, maintenance, entertainment, home education, and communications systems in the house were completely controlled by a central computer. Ms. Parker had shown the students how the combination of a waste-recycling unit, solar cells, and wind turbines met all of the house's energy needs. Ms. Parker had spent a long time talking about the new materials that made such a house possible. Several times, she had drawn pictures of organic and inorganic molecules on the classroom computer screen.

Even with all this preparation, Jennifer did not care about the house. She was interested only in the designs of the past. Jennifer's real love was for twentieth-century antiques. In her opinion, the appliances, furnishings, and materials of 2093 were boring and ugly.

"Jennifer!" Ms. Parker's voice cut through the group.

Jennifer scooped up the bouncing ball and thrust it into her shoulder bag.

"Yes, Ms. Parker?"

"Please come forward and give us the benefit of your views on home design."

As Jennifer moved to the front of the group, she knew she was on the spot. The class was already aware of Jennifer's "good-old-days" views, which contrasted with Ms. Parker's enthusiasm for modern materials and designs.

Nervously, Jennifer started to talk. "Plastics, silicones, polymers...I don't think the scientists of the late twentieth century did us a favor by developing all this phony stuff. I wish we could go back to using wood, steel, aluminum, copper, and real glass. Remember those beautiful steel-and-chrome cars and the polished wood furniture in the design museum?"

"Yes," Ms. Parker agreed, "many things produced in the last century were beautiful and well made. It's easy to admire them in museums. But I don't think we'd find them convenient to live with every day."

"I would," Jennifer insisted.

"All right," responded Ms. Parker, "let's hop into a mental time machine and find out. What age should we go back to?"

"How about the 1990's?" Jeremy suggested.

"Perfect," said Ms. Parker. "The 1990's were just the beginning of the great materials revolution. Great changes in industry, electronics, and materials development began at that time. By the way, Jennifer, you'd better leave your bouncing ball behind. They didn't have elastic glass in the 1990s. Light-

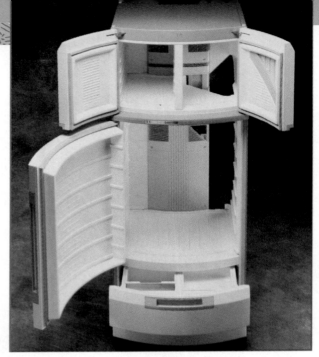

▲ **Plastics are now being used to design technology of the future, such as this model of a refrigerator-freezer.**

sensitive building materials that change color to reduce glare were also unheard of. However, the research to produce these materials was already under way."

A TRIP BACK IN TIME

The house of the future was forgotten as Ms. Parker and the class moved to a nearby park to talk about materials of the past.

Ms. Parker began by asking the students, "What would be the first thing you'd notice in the world of the 1990s?"

"Heaviness. The great weight of almost everything," Jeremy volunteered.

"Go on," Ms. Parker said.

"Appliances such as refrigerators, stoves, washing machines, and air conditioners were still made of metal at that time. They were so heavy that it was almost impossible for one person to lift any of them. And a lot of everyday things were much heavier than they are now. Many food items in supermarkets still came in metal cans and glass bottles. A bagful of those containers could weigh a lot. It wasn't until around 2000 that lightweight, tough plastic containers had completely replaced them."

A SAFER FUTURE

Turning to another student, Ms. Parker asked: "Carlee, what would you notice most about life in the 1990s?"

"That it wasn't safe," Carlee responded. "Why?"

Carlee thought for a moment and then said, "I guess what I was thinking of was the danger of riding in a 1990 car. I've seen pictures of how those old metal cars hurt people in accidents. Sometimes the heavy metal engines and batteries in the front end were pushed back to where people were sitting. Sharp pieces of broken metal and glass were all over the scene of an accident. Cars could blow up or catch fire."

"That can't happen now," Marian said. "Our cars, including the engines, are made of super plastics and silicones inside and out. These materials are very light and strong, and they bounce. Even if a laser brake system fails, no one can get seriously hurt."

"Cars not only are lighter and safer now," Jeremy added, "they require less energy to run. The changeover from metal to plastic engines led to great energy savings. And the changeover since then to solar battery-powered cars has meant even greater savings. If we'd kept on using heavy metal cars, the world might have run out of oil and other materials by now."

"You're quite right, Jeremy," Ms. Parker agreed. "Let's sum up what's happened since the 1990s. A great materials revolution started at that time. Chemists discovered how to produce polymers: very long chains and loops of carbon, oxygen, hydrogen, and nitrogen atoms.

"Around the same time, other scientists created a new family of polymers. Silicon atoms were used instead of carbon atoms, so the materials were called silicones. The result of all this chemistry was a new range of super-strong light, cheap plastics.

"The new materials can be made into anything that one made with metal, wood, glass, or ceramic. In fact, we can do many things with combinations of the new materials that we couldn't do with the old materials, such as making a transparent bouncing ball." Ms. Parker smiled at Jennifer.

Jennifer smiled back, still not convinced of the advantages of living in 2093.

▲ Cars of the future may be made entirely of plastics and powered by solar batteries. Such a car would be lightweight, durable, and clean and inexpensive to operate.

For Further Reading

> If you have been intrigued by the concepts examined in this textbook, you may also be interested in the ways fellow thinkers—novelists, poets, essayists, as well as scientists—have imaginatively explored the same ideas.

Chapter 1: Atoms and Bonding

Bond, Nancy. *The Voyage Begun.* New York: Macmillan.

Botting, Douglas. *The Giant Airships.* New York: Time-Life Books.

Bracy, Norma M. *Salt.* Atwater, CA: Book Binder.

Henry, O. "The Diamond of Kali." In *The Complete Works of O. Henry.* New York: Doubleday.

Lee, Martin. *Paul Revere.* New York: Watts.

Maupassant, Guy de. "The Diamond Necklace." In *The Best Stories of Guy de Maupassant.* New York: Airmont.

Spinelli, Jerry. *Space Station Seventh Grade.* Boston: Little, Brown.

Chapter 2: Chemical Reactions

Anderson, Madelyn K. *Environmental Diseases.* New York: Watts.

Boning, Richard A. *Horror Overhead.* Baldwin, NY: B. Loft.

Burchard, S. H. *The Statue of Liberty: Birth to Rebirth.* New York: Harbrace Junior Books.

Fox, Mary V. *Women Astronauts: Aboard the Space Shuttle.* Englewood Cliffs, NJ: Messner.

Krementz, Jill. *The Fun of Cooking.* New York: Knopf.

Le Guin, Ursula K. *A Wizard of Earthsea.* New York: Bantam.

Stewart, Gail. *Acid Rain.* San Diego, CA: Lucent.

Tchudi, Stephen. *Soda Poppery: The History of Soft Drinks in America.* New York: Macmillan.

Chapter 3: Families of Chemical Compounds

Carter, Alden R. *Up Country.* New York: Putnam.

Dolan, Edward F., Jr. *Great Mysteries of the Ice and Snow.* New York: Putnam.

Fox, Paula. *The Moonlight Man.* New York: Bradbury.

Hendershot, Judith. *In Coal Country.* New York: Knopf.

Kirschmann, John, and Lavon Dunne. *Nutrition Almanac.* New York: McGraw-Hill.

Smith, Robert Kimmel. *Jelly Belly.* New York: Delacorte.

Chapter 4: Petrochemical Technology

Dineen, Jacqueline. *Wool.* Hillside, NJ: Enslow.

Eliot, George. *Silas Marner.* New York: New American Library.

Ferber, Edna. *Giant.* New York: Fawcett.

Merrill, Jean. *The Pushcart War.* New York: Addison-Wesley.

Tarkington, Booth. *The Magnificent Ambersons.* Bloomington, IN: Indiana University Press.

Trier, Mike. *Super Car.* New York: Watts.

Whyman, Kathryn. *Plastics.* New York: Watts.

Chapter 5: Radioactive Elements

Careme, Maurice. *The Peace.* San Marcos, CA: Green Tiger.

Dank, Milton. *Albert Einstein.* New York: Watts.

Hersey, John. *Hiroshima.* New York: Bantam.

Keller, Mollie. *Marie Curie.* New York: Watts.

Mikliowitz, Gloria D. *After the Bomb.* New York: Scholastic.

Nardo, Don. *Chernobyl.* San Diego, CA: Lucent.

Activity Bank

Welcome to the Activity Bank! This is an exciting and enjoyable part of your science textbook. By using the Activity Bank you will have the chance to make a variety of interesting and different observations about science. The best thing about the Activity Bank is that you and your classmates will become the detectives, and as with any investigation you will have to sort through information to find the truth. There will be many twists and turns along the way, some surprises and disappointments too. So always remember to keep an open mind, ask lots of questions, and have fun learning about science.

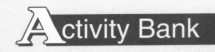
UP IN SMOKE

Electrons surrounding an atomic nucleus are usually located in particular energy levels. Sometimes, however, an electron can absorb extra energy, which forces it up into a higher energy level. An electron in a higher energy level is unstable. Eventually the electron will fall back to its original position, and as it does so, it releases its extra energy in the form of light.

Different elements can absorb and release only certain amounts of energy. The amount of energy determines the color of the light that is given off. Thus the color of light given off can be used to identify particular elements. In this activity you will give extra energy in the form of heat to different elements and observe the colors given off.

Materials You Need

nichrome or platinum wire
cork
Bunsen burner
hydrochloric acid (dilute)
distilled water
7 test tubes
test-tube rack
7 chloride test solutions

Procedure 🔳 🔥 👁

1. Label each of the test tubes with one of the following compounds: LiCl, $CaCl_2$, KCl, $CuCl_2$, $SrCl_2$, NaCl, $BaCl_2$. Pour 5 mL of each test solution into the correctly labeled test tube.

2. Push one end of the wire into the cork and bend the other end into a tiny loop.

3. Put on your safety goggles. Hold the cork and clean the wire by dipping it into the dilute hydrochloric acid and then into distilled water. Then heat the wire in the blue flame of the Bunsen

burner until the wire glows and you can no longer see colors in the flame.

4. Dip the wire into the first test solution. Place the end of the wire at the tip of the inner cone of the burner flame. Record the color given to the flame in a data table similar to the one shown.

5. Clean the wire by repeating step 3. Repeat step 4 for the other 6 test solutions. Make sure you clean the wire after each test.

What You See

DATA TABLE

Compound	Color of Flame
LiCl	
$CaCl_2$	
KCl	
$CuCl_2$	
$SrCl_2$	
NaCl	
$BaCl_2$	

What You Can Conclude

1. Why do you think all of the compounds you tested were bonded to the same element—chlorine?

2. Why did you have to clean the wire before each test?

3. How do your observations compare with those of your classmates?

4. How can you use the flame test to identify a certain element?

HOT STUFF

If you look around, you will find that many of the objects and structures you depend on every day are made of metal. Metals have quite interesting and useful characteristics. One of the most important of these is the fact that metals are excellent conductors of both heat and electricity. In this activity you will find out just how well metals conduct heat.

For this activity, you will need several utensils (spoons or forks) made of different materials, such as silver, stainless steel, plastic, wood, and so forth. You will also need a beaker (or drinking glass), hot water, a pat of frozen butter, and several small objects (beads, frozen peas, popcorn kernels, or raisins).

Press a small glob of butter onto the top of each utensil. Make sure that when the utensils are stood on end, the butter is placed at the same height on each. Be careful not to melt the butter as you work with it. Press a bead, or whatever small object you choose, into the butter. Stand the utensils up in the beaker (leaning on the edge) so that they do not touch each other. Pour hot water into the beaker until it is about 6 cm below the globs of butter.

Watch the utensils for the next several minutes. What do you see happening?

Make a chart listing the material each utensil is made of, and the order in which the bead fell into the water.

Which do you expect to fall first? Which actually does?

Combine your results with those of your classmates. Make a class chart showing all of the materials used and the order in which the beads fell.

THE MILKY WAY

Have you ever squeezed a drop of dishwashing detergent into a pot full of greasy water to watch the grease spread apart? The reason this happens has to do with the molecular structure of both the grease and the detergent. Rather than bonding together, these molecules (or at least parts of them) repel each other and move away. In this activity, you will experiment with a similar example of substances that rearrange themselves when mixed together.

What You Will Need

baking sheet or roasting pan
milk (enough to cover the bottom of the sheet or pan)
food coloring of several different colors
dishwashing detergent

What You Will Do

1. Pour the milk into the baking sheet or pan until the bottom is completely covered.

2. Sprinkle several drops of each different food coloring on the milk. Scatter the drops so that you have drops of different colors all over the milk.

3. Add a few drops of detergent to the middle of the largest blobs of color. What do you see happening?

■ Can you propose a hypothesis to explain your observations?

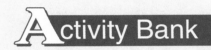
POCKETFUL OF POSIES

Can you picture a meadow filled with wild flowers ranging through all the colors of the rainbow? The beautiful colors of flowers depend on combinations of chemicals carefully selected by Nature. But just as they are formed, they can also be destroyed. In this activity you will create a chemical reaction that affects flower colors.

You will need several flowers of different kinds and colors, a large jar or bottle with a lid (a clean mayonnaise jar or juice bottle will do and a plastic lid is preferable), a rubber band, scissors, and household ammonia (about 50 mL).

Procedure

1. Gather the flowers so that all of the stems are in a bunch. Use the rubber band to hold the stems together. You may have to twist it around the stems more than once.

2. Cut a large hole in the jar lid with the scissors. The gathered stems of the flowers must be able to fit snugly through the hole. **CAUTION:** *Be careful when using sharp instruments.*

3. Push the stems through the hole so that when the lid is placed on the jar, the flowers will be suspended inside the jar.

4. Pour a little ammonia into the jar—enough to cover the bottom. **CAUTION:** *Do not breathe in the ammonia vapors.*

5. Carefully place the lid with the flowers on the jar. Look at the flowers after 20 to 30 minutes. Do you observe any changes in them? If so, what do you see happening?

6. Compare your results with those of your classmates who may have used different flowers. Record the overall results.

Thinking It Through

■ The pigments that give flowers their beautiful colors are present along with chlorophyll, which is green. Chlorophyll is the substance that makes photosynthesis (the food-making process in plants) possible. What do you think happened during this activity to explain your observations?

■ In tree leaves, colorful pigments are also present along with chlorophyll, but in this case the green chlorophyll hides the colorful pigments. What must happen to give leaves their stunning autumn colors?

POPCORN HOP

What do you see when you pour soda or other carbonated drinks into a glass? From experience, you probably know that you see bubbles continually rising to the top—thanks to the carbonation. In this activity you will create a chemical reaction similar to the one occurring in soda and you will use it to make popcorn kernels hop!

Materials

large, clear drinking glass, beaker, or jar
15 mL (1 Tbsp) baking soda
food coloring (a few drops)
45 mL (3 Tbsp) vinegar
popcorn kernels or raisins or mothballs (handful)
stirrer (or long cooking utensil)

Procedure

1. Fill the glass container with water.

2. Add about 15 mL of baking soda, a few drops of food coloring, and stir well.

3. Drop in the popcorn kernels (or raisins/mothballs) and stir in 45 mL of vinegar. Watch the kernels for the next several minutes. What do you observe happening? (If the action slows down, add more baking soda.) Explain what you see in terms of chemical reactions.

■ In an effort to preserve the natural environment, people are beginning to use Earth-friendly cleaning products. For example, rather than dumping poisonous chemicals into a sink, a mixture of hot water, vinegar, and baking soda can be used to clean drains. Why do you think this works?

TOASTING TO GOOD HEALTH

Have you ever been given toast when you weren't feeling well? For some reason, toasted bread seems easier on your digestive system than untoasted bread does. In actuality, the reason is no mystery. It has to do with a chemical reaction involving the heat from your toaster. In this activity you will discover the difference between plain bread and toasted bread.

Materials

slice of white bread
slice of white toasted bread
household iodine (5 mL or 1 tsp)
drinking glass or 250-mL beaker
baking dish (or bowl with a flat bottom)
spoon (measuring spoon would be helpful)

Procedure

1. Fill the drinking glass or beaker half-full with water.

2. Mix 5 mL (about 1 tsp) of iodine into the water. Carefully pour the water-iodine solution into the baking dish.

3. Tear off a strip (about 2 cm wide) from the plain slice of bread. Dip the strip in the solution.

   ■ Do you observe any changes in the bread?

4. Tear off a strip of the same size from the toasted bread. Dip this strip in the solution.

■ Do you observe any changes in the toast?

■ When starch and iodine are combined, they react to form starch iodide, which is a bluish-purple. For this reason, iodine is used to test for starch. Knowing this, what can you learn from your observations?

■ What type of chemical reaction is involved in toasting—endothermic or exothermic?

■ The process of food digestion begins in your mouth. Part of this process involves breaking starches down into simpler substances. As a result of doing this activity, can you now explain why toast is sometimes recommended when you are not feeling well?

IN A JAM

You have probably seen or used litmus paper to determine whether a substance is an acid or a base. But did you know that litmus paper is not the only material that can be used as an acid-base indicator? It may surprise you to learn that many foods can also do the job in a pinch! In this activity you will experiment with just such an indicator.

Materials

blackberry jam (a spoonful is enough)
warm water
small drinking glass
household ammonia (several drops)
lemon juice or vinegar (several drops)
spoon
medicine dropper

Procedure

1. Fill the drinking glass half-full with warm tap water.

2. Put a spoonful of jam into the water and gently stir it with the spoon until it is dissolved. The water-jam solution should turn a reddish color.

3. Use the medicine dropper to put a few drops of ammonia into the solution. Stir the solution once or twice. What happens to the color of the solution?

4. Clean the medicine dropper. Use the clean dropper to add several drops of lemon juice or vinegar to the solution. Clean the spoon and again stir the solution. What happens to the color of the solution this time?

5. Compare your observations with those of your classmates who added the substance that you did not—lemon juice or vinegar. What happened to their solutions?

■ The jam solution is red when an acid is added to it and greenish-purple when a base is added. From your experiment, determine whether ammonia, vinegar, and lemon juice are acids or bases.

The Next Step

Repeat the experiment several more times, each time using a different test substance. You may choose such substances as milk, juice, soda, or fruit. Be sure to clean the spoon between each stirring. Make a chart showing which substances are acids and which are bases. Combine your observations with those of your classmates.

OIL SPILL

You have probably seen television news reports or read newspaper articles about the devastation caused by an oil spill from a supertanker or other holding vessel. But initial reports often underestimate the full spectrum of the damage. In this activity you will simulate interactions with oil so that you can more clearly understand the dangerous consequences of an oil spill.

Materials

medicine dropper
small graduated cylinder
motor oil, used
fan
tongs
3 hard-boiled eggs, not peeled
paper towels
shallow baking pan, about 40 cm × 20 cm
white paper, 1 sheet
graph paper, 1-cm grid
beaker or jar (must be able to hold 3 eggs)

Procedure

1. Partially fill a shallow baking pan two-thirds full with water.

2. Pour the motor oil into the graduated cylinder.

3. Use the medicine dropper to remove 1 mL of oil from the graduated cylinder. Gently squeeze the oil out of the dropper into the center of the pan of water. Describe the interaction between the water and oil.

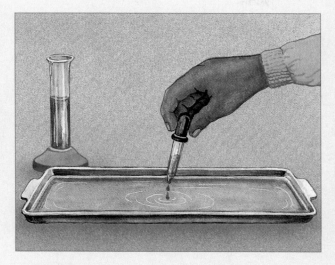

4. Mark off a region on the graph paper that is the same size as the baking pan. After several minutes, sketch the arrangement of oil in the pan of water. When you are finished drawing, count up the number of squares on the graph paper covered by oil. Remember, the area now covered was produced by only 1 mL of oil! Assuming that oil always spreads proportionately, make a chart showing the area that would be covered by 2 mL, 10 mL, 100 mL, and 1 L.

5. Place a fan beside the pan of water and oil. Turn it on and determine if the flow of air affects the spread of oil. What do you discover?

6. Now try shaking the pan slightly. Be careful not to spill the contents. Does this reaction affect the oil at all?

(continued)

7. Gently place the three hard boiled eggs in the jar or beaker. Pour oil into the container until it is full. Place the container under a strong light.

8. After 5 minutes use the tongs to carefully remove one egg. Remove the excess oil with a paper towel. Peel the egg. What do you observe?

9. Remove the second egg after 15 minutes. Peel this egg and record what you observe. Remove the third egg after 30 minutes. Again peel the egg and record your observations.

The Big Picture

1. Supertankers carry millions of liters of oil. In light of your calculations, what can you say about the implications of a large oil spill?

2. What did you learn by blowing air on the oil and by shaking the water? What conditions did these procedures represent? How do these conditions affect the severity of oil spills?

3. What effect could oil have on the eggs of birds nesting near ocean water that becomes contaminated with oil?

THE DOMINO EFFECT

Have you ever played with dominoes? If so, you know that dominoes can be arranged into all sorts of complicated patterns that enable you to knock them all down from a gentle tap on just one domino. Beyond playing, the falling action of dominoes can be used to represent a very complex phenomenon—a nuclear chain reaction. In this activity, you will need 15 dominoes and a stopwatch to learn more about nuclear chemistry.

Procedure

1. Place the dominoes in a row so that each one is standing on its narrow end. Each domino should be about 1–1.5 cm from the next one.

2. Gently tip the first domino in the line over so that it falls on the one behind it. You have just initiated a chain reaction.
   - What keeps the reaction going?
   - How can the row of dominoes be likened to a nuclear chain reaction?

3. Now arrange the dominoes as shown in the accompanying figure.

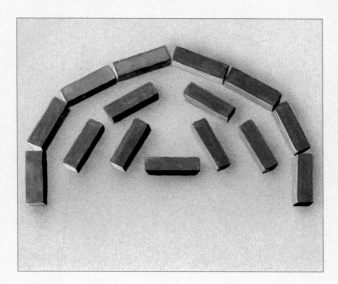

4. Gently tip over the center domino. How does this arrangement differ from the first one?
   - If the dominoes again represent atomic nuclei, how is this chain reaction different from the first one?

5. Which arrangement of dominoes do you think falls faster? Find out. Arrange the dominoes back into a single row. Use the stopwatch to measure the length of time from when you tip the first domino until the last domino falls. Record the measurement. Return the dominoes to the second pattern. Again record the time from the tip of the first domino to the fall of the last one. Which arrangement falls faster?

6. Now set up the dominoes as shown in the accompanying figure.

7. Tip the leftmost domino. Record the amount of time it takes for the dominoes to fall.
   - How does the length of time it takes for the dominoes to fall in this pattern compare with the times for the other two patterns?

(continued)

■ How can you explain this arrangement in terms of a nuclear chain reaction?

8. Design your own arrangement for the dominoes. Determine how long it takes for all the dominoes to fall from this arrangement. Compare this time with those of the other arrangements and with those of your classmates. Make a chart or a poster showing the different arrangements and the length of time recorded for each one.

What This All Means

■ Why might it be important to slow down a reaction?

■ After completing this activity, can you think of how nuclear chain reactions can be controlled in nuclear power plants?

Appendix A

The metric system of measurement is used by scientists throughout the world. It is based on units of ten. Each unit is ten times larger or ten times smaller than the next unit. The most commonly used units of the metric system are given below. After you have finished reading about the metric system, try to put it to use. How tall are you in metrics? What is your mass? What is your normal body temperature in degrees Celsius?

Commonly Used Metric Units

Length   The distance from one point to another

meter (m)   A meter is slightly longer than a yard.
1 meter = 1000 millimeters (mm)
1 meter = 100 centimeters (cm)
1000 meters = 1 kilometer (km)

Volume   The amount of space an object takes up

liter (L)   A liter is slightly more than a quart.
1 liter = 1000 milliliters (mL)

Mass   The amount of matter in an object

gram (g)   A gram has a mass equal to about one paper clip.

1000 grams = 1 kilogram (kg)

Temperature   The measure of hotness or coldness

degrees   0°C = freezing point of water
Celsius (°C)   100°C = boiling point of water

Metric–English Equivalents

2.54 centimeters (cm) = 1 inch (in.)
1 meter (m) = 39.37 inches (in.)
1 kilometer (km) = 0.62 miles (mi)
1 liter (L) = 1.06 quarts (qt)
250 milliliters (mL) = 1 cup (c)
1 kilogram (kg) = 2.2 pounds (lb)
28.3 grams (g) = 1 ounce (oz)
°C = 5/9 × (°F − 32)

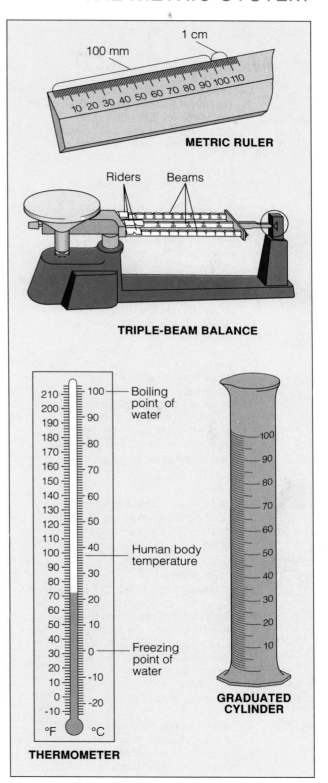

METRIC RULER

TRIPLE-BEAM BALANCE

THERMOMETER

GRADUATED CYLINDER

Glassware Safety

1. Whenever you see this symbol, you will know that you are working with glassware that can easily be broken. Take particular care to handle such glassware safely. And never use broken or chipped glassware.
2. Never heat glassware that is not thoroughly dry. Never pick up any glassware unless you are sure it is not hot. If it is hot, use heat-resistant gloves.
3. Always clean glassware thoroughly before putting it away.

Fire Safety

1. Whenever you see this symbol, you will know that you are working with fire. Never use any source of fire without wearing safety goggles.
2. Never heat anything—particularly chemicals—unless instructed to do so.
3. Never heat anything in a closed container.
4. Never reach across a flame.
5. Always use a clamp, tongs, or heat-resistant gloves to handle hot objects.
6. Always maintain a clean work area, particularly when using a flame.

Heat Safety

Whenever you see this symbol, you will know that you should put on heat-resistant gloves to avoid burning your hands.

Chemical Safety

1. Whenever you see this symbol, you will know that you are working with chemicals that could be hazardous.
2. Never smell any chemical directly from its container. Always use your hand to waft some of the odors from the top of the container toward your nose—and only when instructed to do so.
3. Never mix chemicals unless instructed to do so.
4. Never touch or taste any chemical unless instructed to do so.
5. Keep all lids closed when chemicals are not in use. Dispose of all chemicals as instructed by your teacher.

6. Immediately rinse with water any chemicals, particularly acids, that get on your skin and clothes. Then notify your teacher.

Eye and Face Safety

1. Whenever you see this symbol, you will know that you are performing an experiment in which you must take precautions to protect your eyes and face by wearing safety goggles.
2. When you are heating a test tube or bottle, always point it away from you and others. Chemicals can splash or boil out of a heated test tube.

Sharp Instrument Safety

1. Whenever you see this symbol, you will know that you are working with a sharp instrument.
2. Always use single-edged razors; double-edged razors are too dangerous.
3. Handle any sharp instrument with extreme care. Never cut any material toward you; always cut away from you.
4. Immediately notify your teacher if your skin is cut.

Electrical Safety

1. Whenever you see this symbol, you will know that you are using electricity in the laboratory.
2. Never use long extension cords to plug in any electrical device. Do not plug too many appliances into one socket or you may overload the socket and cause a fire.
3. Never touch an electrical appliance or outlet with wet hands.

Animal Safety

1. Whenever you see this symbol, you will know that you are working with live animals.
2. Do not cause pain, discomfort, or injury to an animal.
3. Follow your teacher's directions when handling animals. Wash your hands thoroughly after handling animals or their cages.

One of the first things a scientist learns is that working in the laboratory can be an exciting experience. But the laboratory can also be quite dangerous if proper safety rules are not followed at all times. To prepare yourself for a safe year in the laboratory, read over the following safety rules. Then read them a second time. Make sure you understand each rule. If you do not, ask your teacher to explain any rules you are unsure of.

Dress Code

1. Many materials in the laboratory can cause eye injury. To protect yourself from possible injury, wear safety goggles whenever you are working with chemicals, burners, or any substance that might get into your eyes. Never wear contact lenses in the laboratory.

2. Wear a laboratory apron or coat whenever you are working with chemicals or heated substances.

3. Tie back long hair to keep it away from any chemicals, burners and candles, or other laboratory equipment.

4. Remove or tie back any article of clothing or jewelry that can hang down and touch chemicals and flames.

General Safety Rules

5. Read all directions for an experiment several times. Follow the directions exactly as they are written. If you are in doubt about any part of the experiment, ask your teacher for assistance.

6. Never perform activities that are not authorized by your teacher. Obtain permission before "experimenting" on your own.

7. Never handle any equipment unless you have specific permission.

8. Take extreme care not to spill any material in the laboratory. If a spill occurs, immediately ask your teacher about the proper cleanup procedure. Never simply pour chemicals or other substances into the sink or trash container.

9. Never eat in the laboratory.

10. Wash your hands before and after each experiment.

First Aid

11. Immediately report all accidents, no matter how minor, to your teacher.

12. Learn what to do in case of specific accidents, such as getting acid in your eyes or on your skin. (Rinse acids from your body with lots of water.)

13. Become aware of the location of the first-aid kit. But your teacher should administer any required first aid due to injury. Or your teacher may send you to the school nurse or call a physician.

14. Know where and how to report an accident or fire. Find out the location of the fire extinguisher, phone, and fire alarm. Keep a list of important phone numbers—such as the fire department and the school nurse—near the phone. Immediately report any fires to your teacher.

Heating and Fire Safety

15. Again, never use a heat source, such as a candle or burner, without wearing safety goggles.

16. Never heat a chemical you are not instructed to heat. A chemical that is harmless when cool may be dangerous when heated.

17. Maintain a clean work area and keep all materials away from flames.

18. Never reach across a flame.

19. Make sure you know how to light a Bunsen burner. (Your teacher will demonstrate the proper procedure for lighting a burner.) If the flame leaps out of a burner toward you, immediately turn off the gas. Do not touch the burner. It may be hot. And never leave a lighted burner unattended!

20. When heating a test tube or bottle, always point it away from you and others. Chemicals can splash or boil out of a heated test tube.

21. Never heat a liquid in a closed container. The expanding gases produced may blow the container apart, injuring you or others.

22. Before picking up a container that has been heated, first hold the back of your hand near it. If you can feel the heat on the back of your hand, the container may be too hot to handle. Use a clamp or tongs when handling hot containers.

Using Chemicals Safely

23. Never mix chemicals for the "fun of it." You might produce a dangerous, possibly explosive substance.

24. Never touch, taste, or smell a chemical unless you are instructed by your teacher to do so. Many chemicals are poisonous. If you are instructed to note the fumes in an experiment, gently wave your hand over the opening of a container and direct the fumes toward your nose. Do not inhale the fumes directly from the container.

25. Use only those chemicals needed in the activity. Keep all lids closed when a chemical is not being used. Notify your teacher whenever chemicals are spilled.

26. Dispose of all chemicals as instructed by your teacher. To avoid contamination, never return chemicals to their original containers.

27. Be extra careful when working with acids or bases. Pour such chemicals over the sink, not over your workbench.

28. When diluting an acid, pour the acid into water. Never pour water into an acid.

29. Immediately rinse with water any acids that get on your skin or clothing. Then notify your teacher of any acid spill.

Using Glassware Safely

30. Never force glass tubing into a rubber stopper. A turning motion and lubricant will be helpful when inserting glass tubing into rubber stoppers or rubber tubing. Your teacher will demonstrate the proper way to insert glass tubing.

31. Never heat glassware that is not thoroughly dry. Use a wire screen to protect glassware from any flame.

32. Keep in mind that hot glassware will not appear hot. Never pick up glassware without first checking to see if it is hot. See #22.

33. If you are instructed to cut glass tubing, fire-polish the ends immediately to remove sharp edges.

34. Never use broken or chipped glassware. If glassware breaks, notify your teacher and dispose of the glassware in the proper trash container.

35. Never eat or drink from laboratory glassware. Thoroughly clean glassware before putting it away.

Using Sharp Instruments

36. Handle scalpels or razor blades with extreme care. Never cut material toward you; cut away from you.

37. Immediately notify your teacher if you cut your skin when working in the laboratory.

Animal Safety

38. No experiments that will cause pain, discomfort, or harm to mammals, birds, reptiles, fishes, and amphibians should be done in the classroom or at home.

39. Animals should be handled only if necessary. If an animal is excited or frightened, pregnant, feeding, or with its young, special handling is required.

40. Your teacher will instruct you as to how to handle each animal species that may be brought into the classroom.

41. Clean your hands thoroughly after handling animals or the cage containing animals.

End-of-Experiment Rules

42. After an experiment has been completed, clean up your work area and return all equipment to its proper place.

43. Wash your hands after every experiment.

44. Turn off all burners before leaving the laboratory. Check that the gas line leading to the burner is off as well.

Appendix D

NAME	SYMBOL	ATOMIC NUMBER	ATOMIC MASS†	NAME	SYMBOL	ATOMIC NUMBER	ATOMIC MASS†
Actinium	Ac	89	(227)	Neodymium	Nd	60	144.2
Aluminum	Al	13	27.0	Neon	Ne	10	20.2
Americium	Am	95	(243)	Neptunium	Np	93	(237)
Antimony	Sb	51	121.8	Nickel	Ni	28	58.7
Argon	Ar	18	39.9	Niobium	Nb	41	92.9
Arsenic	As	33	74.9	Nitrogen	N	7	14.01
Astatine	At	85	(210)	Nobelium	No	102	(255)
Barium	Ba	56	137.3	Osmium	Os	76	190.2
Berkelium	Bk	97	(247)	Oxygen	O	8	16.00
Beryllium	Be	4	9.01	Palladium	Pd	46	106.4
Bismuth	Bi	83	209.0	Phosphorus	P	15	31.0
Boron	B	5	10.8	Platinum	Pt	78	195.1
Bromine	Br	35	79.9	Plutonium	Pu	94	(244)
Cadmium	Cd	48	112.4	Polonium	Po	84	(210)
Calcium	Ca	20	40.1	Potassium	K	19	39.1
Californium	Cf	98	(251)	Praseodymium	Pr	59	140.9
Carbon	C	6	12.01	Promethium	Pm	61	(145)
Cerium	Ce	58	140.1	Protactinium	Pa	91	(231)
Cesium	Cs	55	132.9	Radium	Ra	88	(226)
Chlorine	Cl	17	35.5	Radon	Rn	86	(222)
Chromium	Cr	24	52.0	Rhenium	Re	75	186.2
Cobalt	Co	27	58.9	Rhodium	Rh	45	102.9
Copper	Cu	29	63.5	Rubidium	Rb	37	85.5
Curium	Cm	96	(247)	Ruthenium	Ru	44	101.1
Dysprosium	Dy	66	162.5	Samarium	Sm	62	150.4
Einsteinium	Es	99	(254)	Scandium	Sc	21	45.0
Erbium	Er	68	167.3	Selenium	Se	34	79.0
Europium	Eu	63	152.0	Silicon	Si	14	28.1
Fermium	Fm	100	(257)	Silver	Ag	47	107.9
Fluorine	F	9	19.0	Sodium	Na	11	23.0
Francium	Fr	87	(223)	Strontium	Sr	38	87.6
Gadolinium	Gd	64	157.2	Sulfur	S	16	32.1
Gallium	Ga	31	69.7	Tantalum	Ta	73	180.9
Germanium	Ge	32	72.6	Technetium	Tc	43	(97)
Gold	Au	79	197.0	Tellurium	Te	52	127.6
Hafnium	Hf	72	178.5	Terbium	Tb	65	158.9
Helium	He	2	4.00	Thallium	Tl	81	204.4
Holmium	Ho	67	164.9	Thorium	Th	90	232.0
Hydrogen	H	1	1.008	Thulium	Tm	69	168.9
Indium	In	49	114.8	Tin	Sn	50	118.7
Iodine	I	53	126.9	Titanium	Ti	22	47.9
Iridium	Ir	77	192.2	Tungsten	W	74	183.9
Iron	Fe	26	55.8	Unnilennium	Une	109	(266?)
Krypton	Kr	36	83.8	Unnilhexium	Unh	106	(263)
Lanthanum	La	57	138.9	Unniloctium	Uno	108	(265)
Lawrencium	Lr	103	(256)	Unnilpentium	Unp	105	(262)
Lead	Pb	82	207.2	Unnilquadium	Unq	104	(261)
Lithium	Li	3	6.94	Unnilseptium	Uns	107	(262)
Lutetium	Lu	71	175.0	Uranium	U	92	238.0
Magnesium	Mg	12	24.3	Vanadium	V	23	50.9
Manganese	Mn	25	54.9	Xenon	Xe	54	131.3
Mendelevium	Md	101	(258)	Ytterbium	Yb	70	173.0
Mercury	Hg	80	200.6	Yttrium	Y	39	88.9
Molybdenum	Mo	42	95.9	Zinc	Zn	30	65.4
				Zirconium	Zr	40	91.2

†Numbers in parentheses give the mass number of the most stable isotope.

Key

6	Atomic number
C	Element's symbol
Carbon	Element's name
12.011	Atomic mass

1

1

1
H
Hydrogen
1.00794

2

2

3	4
Li	**Be**
Lithium	Beryllium
6.941	9.0122

3

11	12
Na	**Mg**
Sodium	Magnesium
22.990	24.305

Transition Metals

	3	**4**	**5**	**6**	**7**	**8**	**9**

4

19	20	21	22	23	24	25	26	27
K	**Ca**	**Sc**	**Ti**	**V**	**Cr**	**Mn**	**Fe**	**Co**
Potassium	Calcium	Scandium	Titanium	Vanadium	Chromium	Manganese	Iron	Cobalt
39.098	40.08	44.956	47.88	50.94	51.996	54.938	55.847	58.9332

5

37	38	39	40	41	42	43	44	45
Rb	**Sr**	**Y**	**Zr**	**Nb**	**Mo**	**Tc**	**Ru**	**Rh**
Rubidium	Strontium	Yttrium	Zirconium	Niobium	Molybdenum	Technetium	Ruthenium	Rhodium
85.468	87.62	88.9059	91.224	92.91	95.94	(98)	101.07	102.906

6

55	56	57 to 71	72	73	74	75	76	77
Cs	**Ba**		**Hf**	**Ta**	**W**	**Re**	**Os**	**Ir**
Cesium	Barium		Hafnium	Tantalum	Tungsten	Rhenium	Osmium	Iridium
132.91	137.33		178.49	180.95	183.85	186.207	190.2	192.22

7

87	88	89 to 103	104	105	106	107	108	109
Fr	**Ra**		**Unq**	**Unp**	**Unh**	**Uns**	**Uno**	**Une**
Francium	Radium		Unnilquadium	Unnilpentium	Unnilhexium	Unnilseptium	Unniloctium	Unnilennium
(223)	226.025		(261)	(262)	(263)	(262)	(265)	(266)

Rare-Earth Elements

Lanthanoid Series

57	58	59	60	61	62
La	**Ce**	**Pr**	**Nd**	**Pm**	**Sm**
Lanthanum	Cerium	Praseodymium	Neodymium	Promethium	Samarium
138.906	140.12	140.908	144.24	(145)	150.36

Actinoid Series

89	90	91	92	93	94
Ac	**Th**	**Pa**	**U**	**Np**	**Pu**
Actinium	Thorium	Protactinium	Uranium	Neptunium	Plutonium
227.028	232.038	231.036	238.029	237.048	(244)

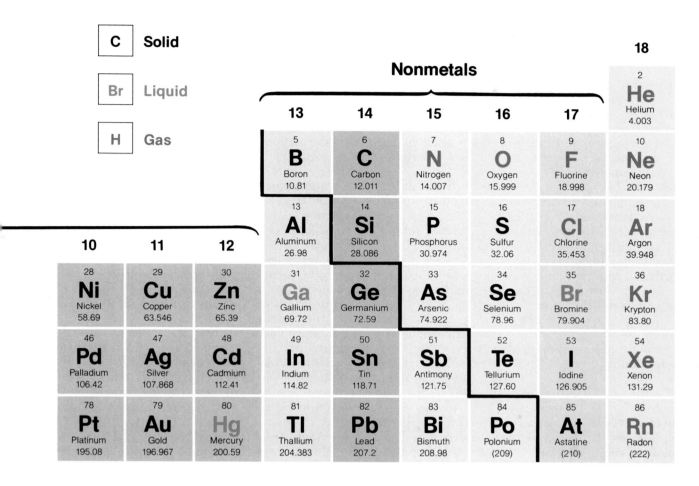

The symbols shown here for elements 104-109 are being used temporarily until names for these elements can be agreed upon.

Metals

Mass numbers in parentheses are those of the most stable or common isotope.

| 63 Eu Europium 151.96 | 64 Gd Gadolinium 157.25 | 65 Tb Terbium 158.925 | 66 Dy Dysprosium 162.50 | 67 Ho Holmium 164.93 | 68 Er Erbium 167.26 | 69 Tm Thulium 168.934 | 70 Yb Ytterbium 173.04 | 71 Lu Lutetium 174.967 |
| 95 Am Americium (243) | 96 Cm Curium (247) | 97 Bk Berkelium (247) | 98 Cf Californium (251) | 99 Es Einsteinium (252) | 100 Fm Fermium (257) | 101 Md Mendelevium (258) | 102 No Nobelium (259) | 103 Lr Lawrencium (260) |

Glossary

acid: compound with a pH below 7 that tastes sour, turns blue litmus paper red, reacts with metals to produce hydrogen gas, and ionizes in water to produce hydrogen ions; proton donor

activation energy: energy required for a chemical reaction to occur

alkane: straight-chain or branched-chain saturated hydrocarbon

alkene: unsaturated hydrocarbon in which at least one pair of carbon atoms is joined by a double covalent bond

alkyne: unsaturated hydrocarbon in which at least one pair of carbon atoms is joined by a triple covalent bond

alpha (AL-fuh) **particle:** weakest type of nuclear radiation; consists of a helium nucleus released during alpha decay

artificial transmutation: changing of one element into another by unnatural means; involves bombarding a nucleus with high-energy particles to cause change

atom: smallest part of any element that retains all the properties of that element

base: compound with pH above 7 that tastes bitter, is slippery to the touch, turns red litmus paper blue, and ionizes in water to produce hydroxide ions; proton acceptor

beta (BAYT-uh) **particle:** electron, created in the nucleus of an atom, released during beta decay

binding energy: energy associated with the strong nuclear force that holds an atomic nucleus together; related to the stability of a nucleus

bubble chamber: device that uses a superheated liquid to create bubbles when radioactive particles pass through it

catalyst: substance that increases the rate of a chemical reaction without being changed by the reaction

chemical bonding: combining of atoms of elements to form new substances

chemical equation: description of a chemical reaction using symbols to represent elements and formulas to represent equations

chemical reaction: process in which substances undergo physical and chemical changes that result in the formation of new substances with different properties

cloud chamber: device to study radioactivity, which uses a cooled gas that will condense around radioactive particles

collision theory: theory that relates collisions among particles to reaction rate; reaction rate depends on such factors as concentration, surface area, temperature, and catalysts

concentrated solution: solution in which a large amount of solute is dissolved in a solvent

concentration: amount of a solute dissolved in a certain amount of solvent

covalent bonding: bonding that involves the sharing of electrons

crystal lattice: regular, repeating arrangement of atoms

decay series: sequence of steps by which a radioactive nucleus decays into a nonradioactive nucleus

decomposition reaction: chemical reaction in which a complex substance breaks down into two or more simpler substances

diatomic element: element whose atoms can form covalent bonds with another atom of the same element

dilute solution: solution in which there is only a little dissolved solute

double-replacement reaction: chemical reaction in which different atoms in two different compounds replace each other

electrolyte (ee-LEHK-troh-light): substance whose water solution conducts electric current

electron affinity: tendency of an atom to attract electrons

electron-dot diagram: diagram that uses the chemical symbol for an element surrounded by a series of dots to represent the electron sharing that takes place in a covalent bond

electroscope: device consisting of a metal rod with two thin metal leaves at one end that can be used to detect radioactivity

endothermic reaction: chemical reaction in which energy is absorbed

exothermic (ek-soh-THER-mihk) **reaction:** chemical reaction in which energy is released

fraction: petroleum part with its own boiling point

gamma (GAM-uh) **ray:** high-frequency electromagnetic wave released during gamma decay; strongest type of nuclear radiation

Geiger counter: device that can be used to detect radioactivity because it produces an electric current in the presence of a radioactive substance

half-life: amount of time it takes for half the atoms in a given sample of an element to decay

hydrocarbon: organic compound that contains only hydrogen and carbon

ion: an atom that has become charged due to the loss or gain of electrons

ionic bonding: bonding that involves the transfer of electrons

ionization: process of removing electrons and forming ions

ionization energy: energy required for ionization

isomer: one of a number of compounds that have the same molecular formula but different structures

isotope (IGH-suh-tohp): atom that has the same number of protons (atomic number) as another atom but a different number of neutrons

kinetics: study of the rates of chemical reactions

metallic bond: bond formed by atoms of metals, in which the outer electrons of the atoms form a common electron cloud

molecule (MAHL-ih-kyool): combination of atoms formed by a covalent bond

monomer: smaller molecule that joins with other smaller molecules to form a chain molecule called a polymer

natural polymer: polymer molecule found in nature; for example, cotton, silk, and wool

network solid: covalent substance whose molecules are very large because the atoms involved continue to bond to one another; have rather high melting points

neutralization (noo-truhl-ih-ZAY-shun): reaction in which an acid combines with a base to form a salt and water

nonelectrolyte: substance whose water solution does not conduct electric current

nuclear chain reaction: series of fission reactions that occur because the products released during one fission reaction cause fission reactions in other atoms

nuclear fission (FIHSH-uhn): splitting of an atomic nucleus into two smaller nuclei of approximately equal mass

nuclear fusion (FYOO-zhuhn): joining of two atomic nuclei of smaller mass to form a single nucleus of larger mass

nuclear radiation: particles and energy released from a radioactive nucleus

nuclear strong force: force that overcomes the electric force of repulsion among protons in an atomic nucleus and binds the nucleus together

organic compound: compound that contains carbon

oxidation number: number of electrons an atom gains, loses, or shares when it forms chemical bonds

petrochemical product: product made either directly or indirectly from petroleum

petroleum: substance believed to have been formed hundreds of millions of years ago when dead plants and animals were buried beneath sediments such as mud, sand, silt, or clay at the bottom of the oceans; crude oil

pH: measure of the hydronium ion concentration of a solution; measured on a scale from 0 to 14

polyatomic ion: group of covalently bonded atoms that acts like a single atom when combining with other atoms

polymer: large molecule in the form of a chain whose links are smaller molecules called monomers

polymerization (poh-lihm-er-uh-ZAY-shuhn): process of chemically bonding monomers to form polymers

product: substance produced by a chemical reaction

radioactive: description for a nucleus that gives off nuclear radiation in the form of mass and energy in order to become stable

radioactive decay: process in which a nucleus spontaneously emits particles or rays to become lighter and more stable

radioactivity: release of energy and matter that results from changes in the nucleus of an atom

radioisotope: artificially produced radioactive isotope; often used in medicine or industry

reactant (ree-AK-tuhnt): substance that enters into a chemical reaction

reaction rate: measure of how quickly reactants change into products

refining: process of separating petroleum into its fractions

salt: compound formed from the positive ion of a base and the negative ion of an acid

saturated hydrocarbon: hydrocarbon in which all the bonds between carbon atoms are single covalent bonds

saturated solution: solution that contains all the solute it can hold at a given temperature

single-replacement reaction: chemical reaction in which an uncombined element replaces an element that is part of a compound

solubility (sahl-yoo-BIHL-uh-tee): measure of how much of a solute can be dissolved in a given amount of solvent under certain conditions

solute (SAHL-yoot): substance that is dissolved in a solution

solution: mixture in which one substance is dissolved, or broken down, in another substance

solvent (SAHL-vuhnt): substance in a solution that does the dissolving

structural formula: description of a molecule that shows the kind, number, and arrangement of atoms in a molecule

substituted hydrocarbon: hydrocarbon formed when one or more hydrogen atoms in a hydrocarbon ring or chain is replaced by a different atom or group of atoms

supersaturated solution: unstable solution that holds more solute than is normal for a given temperature

synthesis (SIHN-thuh-sihs) **reaction:** chemical reaction in which two or more simple substances combine to form a new, more complex substance

synthetic polymer: polymer that does not occur naturally; formed from petrochemicals by people

tracer: radioactive element whose pathway can be followed through the steps of a chemical reaction or industrial process

transmutation: process in which one element is changed into another as a result of changes in the nucleus

transuranium element: element formed synthetically; has more than 92 protons in its nucleus

unsaturated hydrocarbon: hydrocarbon in which one or more of the bonds between carbon atoms is a double covalent or triple covalent bond

unsaturated solution: solution that contains less solute than it can possibly hold at a given temperature

valence electron: electron in the outermost energy level of an atom

Index